T0237132

ENGINES
AN INTRODUCTION

The internal combustion engine that powers the modern automobile has changed very little from its initial design of some eighty years ago. Unlike many high tech advances, engine design still depends on an understanding of basic fluid mechanics and thermodynamics.

This text offers a fresh approach to the study of engines, with an emphasis on design and on fluid dynamics, a current research area in engine design. Professor Lumley, a renowned fluid dynamicist, provides a lucid explanation of how air and fuel are mixed, how they get into the engine, what happens to them there, and how they get out again. Particular attention is given to the complex issue of pollution. Every chapter includes numerous illustrations and examples and concludes with homework problems. Lending a historical note, design sections begin with the earliest example of the design being described. The latest designs are also treated, including stratified charge gasoline direct injection engines.

It is intended that the text be used in conjunction with the Stanford Engine Simulation Program (ESP). ESP is a user-friendly, interactive program modeling flow into and out of one cylinder of a multi-cylinder engine. The thermodynamic analysis assumes a homogeneous charge, a simple geometric approach to flame structure, and a one-equation dynamical turbulence model that allows the effect of turbulence on heat transfer and combustion to be examined. This software tool answers a significant need that is unaddressed by other texts on engines.

Aimed at undergraduate and first-year graduate students, the book will also appeal to hobbyists and car buffs who will appreciate the wealth of illustrations of classic, racing, and modern engines.

John L. Lumley is the Willis H. Carrier Professor of Engineering, Sibley School of Mechanical and Aerospace Engineering, Cornell University. He is a Member of the National Academy of Engineering and a Fellow of the American Academy of Arts and Sciences.

ENGINES

AN INTRODUCTION

JOHN L. LUMLEY

Sibley School of Mechanical and Aerospace Engineering, Cornell University

USING VERSION 2
OF THE STANFORD ENGINE
SIMULATION PROGRAM
OF W. C. REYNOLDS

CAMBRIDGE
UNIVERSITY PRESS

CAMBRIDGE UNIVERSITY PRESS
Cambridge, New York, Melbourne, Madrid, Cape Town, Singapore,
São Paulo, Delhi, Dubai, Tokyo, Mexico City

Cambridge University Press
32 Avenue of the Americas, New York, NY 10013-2473, USA

www.cambridge.org
Information on this title: www.cambridge.org/9780521644891

© Cambridge University Press 1999

This publication is in copyright. Subject to statutory exception
and to the provisions of relevant collective licensing agreements,
no reproduction of any part may take place without the written
permission of Cambridge University Press.

First published 1999
Reprinted 2000, 2004, 2006, 2008, 2009

A catalog record for this publication is available from the British Library

Library of Congress Cataloging in Publication data
Lumley, John L. (John Leask), 1930–
    Engines : an introduction / John L. Lumley.
        p.   cm.
    ISBN 0-521-64277-9 (hardback). – ISBN 0-521-64489-5 (pbk.)
    1. Internal combustion engines.   I. Title.
TJ755.L94   1999
621.43 – dc21
                                                      99-11974
                                                        CIP

ISBN 978-0-521-64277-4 Hardback
ISBN 978-0-521-64489-1 Paperback

Cambridge University Press has no responsibility for the persistence or
accuracy of URLs for external or third-party internet websites referred to in
this publication, and does not guarantee that any content on such websites is,
or will remain, accurate or appropriate. Information regarding prices, travel
timetables, and other factual information given in this work are correct at
the time of first printing but Cambridge University Press does not guarantee
the accuracy of such information thereafter.

I dedicate this book to my father, who got me interested in cars;
to Samuel Beeler, who taught me a lot about them;
and to my grandchildren, who are showing signs of interest.
I hope there will be some mystique left for them.

CONTENTS

PREFACE

As soon as the steam engine was developed in the early 19th century, people started to think about some sort of road vehicle. Early attempts were steam powered and had various engine and boiler positions and arrangements of road wheels. The development of the much more compact spark-ignition internal-combustion engine toward the end of the century made possible a horseless carriage configuration. By about 1910, the automobile had taken a recognizably modern shape. The engine, in particular, has not changed much in basic design since that period. Of course, there are various forms of engine (for example, various arrangements of the valves), having different characteristics. These different forms represented innovations when first proposed. Some of them stood the test of time, and the unsuccessful ones disappeared without a trace. Some uncommon ones (for example, desmodromic valves, which we shall discuss later) keep reappearing from time to time. There is very little new in engines today. For example, around the time of the First World War the enthusiast driver could buy a Frontenac aftermarket cylinder head conversion for his Model T Ford, giving it chain-driven double overhead cams (DOHC) with four valves per cylinder. These are innovations we think of as modern. In [3] we find the statement "OHC engines are more efficient than their predecessor pushrod... engines," which indicates that it was written by a young person (although to be fair we must admit that DOHC engines with four valves per cylinder at first appeared largely in racing and exotic cars). The Fronty conversions were very popular on the dirt-track racing circuit.

There have been changes in the engine since the teens of the century, but these have been slow changes in engine proportions, and changes in materials, rather than changes in basic design. We will discuss later the changes in stroke/bore ratio; early engines had a fairly long stroke relative to the bore, and slowly over the years the stroke has shortened, until it is now usually of the same order as the bore. We will discover later that a shorter stroke allows a higher power for the same displacement, at a higher engine speed, so there has been a progressive change to smaller displacement engines that achieve their peak power at higher speeds. There are very good reasons why the present stroke/bore ratio is desirable; what is not clear is why the earlier engine designers preferred the larger ratio. Of course, automobile design is not always entirely rational. For example, until 1923 essentially all racing engines had four valves per cylinder, which is a very sensible thing to have, as we shall see. However, after 1923 they changed to two valves per cylinder. The reason for this was the achievement of a Fiat racing engine, which exceeded the then-magic figure of 50 brake horsepower per liter (now 37 kW/L) for the first time that year (this is still a good figure for a normal street automobile, although high performance cars double that figure now) with only two valves per cylinder. We will see later that the two things (the number of valves, and the specific output) were in this case completely unrelated (in general the relation goes the other way – more valves produces higher specific output), but everybody changed to two valves per cylinder under the assumption that Fiat must have known something. The automobile industry, and individual designers, are terribly conservative; the design of a new engine, and the construction of a prototype, is a very expensive business, and costs increase steeply when failure and redesign are part of the scenario.

Changes in materials have been significant. Before the Second World War, the useful life of an engine was in the neighborhood of 70,000 miles. At this point one could normally expect that the cylinders would have taper and out-of-round of perhaps three-thousandths of an inch or more, necessitating a rebore, and that the crank throws and mains would have several thousandths ovality, requiring a regrind. Now, engines last at least 100,000 miles, and engines in small and large trucks much more. To a certain extent, the lifetime of an engine is designed into it. A pickup truck owner will be upset if his engine does not last between 150,000 and 250,000 miles, and the owner of an 18-wheeler probably expects half-a-million miles. The increased design lifetime is accomplished by careful attention to keeping piston speed down, bearing loads low, and valve- and piston-crown temperatures low. This means that the engine will be larger, heavier and more expensive for the same power output, but in a truck that is usually not a major consideration. Partly, the increased lifetime results from vast improvements in air cleaners, preventing airborne grit from entering the cylinders and the oil. Partly, it is due to improvements in metallurgy, particularly cylinder walls, piston rings, valves and valve seats, and in the design of thin-shell bearings. Thin-shell bearings can carry much higher unit loads than the old thick, poured babbit bearings, which cracked and shed bits and pieces into the oil when over stressed. There have been great advances in lubrication. Before the Second World War, oils broke down rapidly at engine operating temperatures, and the resulting thick black sludge clogged everything, preventing proper lubrication. Rebuilding an engine was a very messy business. Oils now are far more resistant to breakdown, and detergent oils keep impurities in suspension, so that they leave the engine when the oil is changed. In addition, multi-grade oils provide good lubrication over a much wider range of temperatures. Finally, the development of plastics and synthetic rubbers around the time of the Second World War has made for much better seals, resulting in less leakage of vital fluids, and making possible simpler design. The timing of all these developments (metallurgy, lubrication, rubbers and plastics) coincides with the Second World War, because they were developed in wartime for the Allied tanks, trucks and aircraft, in some cases in the face of unavailability of previously used materials.

The automobile, and its engine, are much more than technology: they occupy a unique sociocultural niche in our lives. Even the most draconian governmental measures (such as punishing European taxes on gasoline and oil) cannot make people take public transport; only a total inability to park in clogged city centers can accomplish that. The automobile has given us at least the illusion of liberation. The feeling is probably not new; a young man in the middle ages probably felt just as liberated by the posession of a spirited horse. Automobiles have an added advantage for American youth, who carry out their courtship in them (or they did when I was young); it was the only place they could really call their own. Some people live in their cars. The automobile brings a feeling of power, agility and speed even to the klutzes among us; a French driver on his way from Marseille to Strasbourg for a weekend of hunting imagines himself on the Circuit Paul Ricard.

Initially, the automobile was a masculine domain, as was nearly everything else outside the home. However, beginning at least in the 1920s, district nurses and other professional women bought Model T or A Fords to cover larger districts and generally give themselves greater mobility, and thus enjoyed the same feeling of independence and empowerment as the men, and got the same kick out of rushing down the country lanes and scaring the chickens. Now, of course, we have notable women race drivers, engineers, mechanics, enthusiasts and aggressive commuters.

Knowledge of the automobile, and in particular the engine, has become the measure of a macho person, man or woman. In bars and on golf courses they discuss 0–60 times, number of valves per cylinder, number of horsepower per liter and similar statistics. For a segment of the population these replace baseball statistics. There is certainly a lot of nonsense talked, and the numbers of most interest in the bar are not necessarily those of greatest interest to an engine designer. In part, this kind of interest arises because the car becomes an extension of the personality. If you were an enthusiast, you could envision fitting your car with high-compression heads, a high-lift cam, multiple Weber carbs, possibly a turbocharger, and many other interesting things to modify the performance, creating for yourself a personalized sports car. The suspension and body could be similarly modified. More creative owners, with access to a machine shop, could envision modifications beyond off-the-shelf aftermarket parts. When I was in high school in Detroit, my friends and I were doing this kind of thing. That was the era of hot rods and lead sleds.[1] Although relatively few owners actually modified their cars in serious ways, a much larger number dreamed of it, and it was at least a theoretical possibility for an even larger number, who could realistically afford no more than fuzzy dice to hang from the rear view mirror. Even the *idea* that you might modify your car to conform to your fantasies meant that you and the car had a very special relationship. This kind of personalization is increasingly difficult because engines (and cars) are more and more dependent on computer control, which is usually beyond the modest capabilities of the shade tree mechanic. At the very least, however, you can still buy a model to fit your fantasy, your vision of yourself.

---

[1] A lead sled is a car that has had all the body seams welded and filled with body solder to create a visual effect like the Volkswagen Karman Ghia.

This book assumes that the reader will have access to the Stanford Engine Simulation Program, developed by W. C. Reynolds at Stanford University, now available as freeware on the ESP website http://esp.stanford.edu. The physics of this program is outlined in Chapter 8. Briefly, ESP is a fast-running, flexible, user-friendly interactive program, for simulation of the thermodynamic performance of homogeneous-charge engines. A single cylinder is considered using a thermodynamic analysis which assumes that the charge is homogeneous, a simple geometric approach to flame structure, and a one-equation dynamical turbulence model allowing examination of the effects of turbulence on heat transfer and combustion. In addition, the program simulates the flow in the intake and exhaust manifolds, permitting studies of manifold tuning.

The user specifies geometric parameters of the engine, including bore, stroke, rod length, valve lift and timing, the heat transfer area above the piston at top center, number of cylinders, and dimensions and parameters specifying the various parts of the manifolds. The user also determines operating parameters including engine speed, spark timing and manifold pressures. The parameters can be adjusted to get reasonable agreement with actual engine data, and the model then used to study the effects of proposed design changes.

In addition, the programs STANJAN and ESPJAN are available at the same website, which permit preparation of tables of thermodynamic properties of the fresh charge and the products for loading into ESP.[2]

I hope the reader will use ESP to experiment with the influence of the various design changes that are discussed in this book.

This book grew out of a course in Automotive Engineering that I teach at Cornell University. The material on engines occupies the first five weeks of the course, but there is certainly more than five weeks' material in this book. The course is intended for engineering students in their third year, although it is also occasionally taken by students in their fourth and fifth years. Students in their fourth and fifth years have to complete several design projects. The entire course, and certainly the section on engines, has a strong orientation toward design. In the section on engines, I try to address questions such as: should an engine have two, three, four or five valves per cylinder? Overhead cams or pushrods? Flat head, hemi head? Squish? Four, five, six, eight, twelve, or sixteen cylinders? Some questions are of a very fundamental nature, such as how can we compare on a more or less equal footing the design of very different engines?

This book is not intended solely for these students, however. I have restored old cars for fifty years, and have a connoisseur's interest in old machinery. As I have indicated, the gasoline engine is old technology, perhaps 80 years old. I hope this book will also be interesting to the enthusiast and to the old car buff.

---

[2] As a historical note, JAN is short for JANAF, which stands for Joint Army-Navy-Air Force (Ad-Hoc Panel on Performance Calculation Methods and Thermodynamic Data, set up in 1958). The JANAF Thermochemical Tables in hard-copy format were for many years the ultimate source of thermochemical data. STANJAN, a program developed at Stanford by W. C. Reynolds, uses these tables as the basic data in a unique element potential method for chemical equilibrium analysis.

I have tried to draw illustrations from some of the wonderful street engines of the past half century, many of which can still be seen at car shows, as well as from racing engines from the same period. However, I have not neglected modern engines; the comparisons are interesting, and we will find they are often not much changed from the earlier engines. I hope that an enthusiast or car buff will enjoy this material and will learn enough about engines to be able to evaluate for himself the design of some of his favorite engines.

My particular area of research expertise is fluid mechanics, especially turbulence; that is, how fluids flow through and around objects, particularly when the flow is irregular and unsteady. As a result, I have a lot to say about how the air and fuel are mixed, how they get into the engine, what happens to them there, and how the products get out again. There are also interesting fluid-mechanical things going on in the crankcase and in the cooling jacket. I have tried to give particular attention to the fluid mechanics of the engine in this book. (The aerodynamics of the vehicle exterior is also fascinating, but is not within the scope of this work – see [50].)

ACKNOWLEDGMENTS

This book grew out of notes for a course in Automotive Engineering which I have taught since 1992 at Cornell University. The course was started by Albert R. George in 1981. When I took it over, Al generously loaned me all the materials I needed to teach it; despite the changes and additions I have made, the structure of this book necessarily still bears a strong family resemblance to Al's course. I am very grateful for his help, and hope that he will forgive me for the rather idiosyncratic tilt I have given the material, for which I take full responsibility.

The Stanford Engine Simulation Program was made available to me by Bill Reynolds. He worked intensively on the program, extending it at my suggestion to include an original simplified manifold dynamic model, and modifying it to improve comparison with experiment. He was extremely generous with his time as well as his program and was in addition the soul of patience in our collaboration. I thank him.

The entire manuscript was carefully read by Chris Edwards, of Stanford University, and by the students in his engines course, all of whom made extensive comments which I found very helpful.

The book was used in manuscript form for the engines course at Ohio State University in the spring of 1998 by Yann Guezennec, which also produced helpful comments.

Chapter 5 was read with great care by Dan Haworth of the General Motors Research Laboratory, who made extensive and very helpful comments for which I am grateful. Julian Hunt of Cambridge University and Jacques Borée of The Institut de Mécanique des Fluides de Toulouse also were kind enough to provide material and commentary that influenced Chapter 5.

The fluid mechanics embodied in this book spring from research which has been supported variously and most notably over the years by the U.S. Office of Naval Research, the Air Force Office of Scientific Research, the U.S. National Science Foundation, and the National Aeronautics and Space Administration. I appreciate their generous and unwavering support.

I would like to point out here that this book is an example of the kind of unplanned benefit to our national practical capabilities that can, in fact, be expected to arise from generous public support of basic research.

ENGINES
AN INTRODUCTION

1
THERMODYNAMIC CONSIDERATIONS

1.1
THE IDEAL OTTO CYCLE

A *cycle* is an idealization of what goes on in one of the devices that thermodynamicists call heat engines: that is, a gasoline or diesel engine, a jet engine, a steam engine, and so forth. All of these take some energy source and convert some of that energy into useful work. In the spark-ignition engine the energy source is a chemical fuel, usually gasoline, which is combined with oxygen from the air by burning to release heat. Expansion of the heated gases does the mechanical work.

For the spark-ignition engine the idealization is called the Otto cycle, after Dr. N. A. Otto who, in 1876, patented a stationary gas engine using approximately this cycle. In order to understand this ideal cycle, we must imagine a piston in a cylinder. The piston is connected to a crank by a connecting rod – see Figure 1.1. The crank rotates, and the piston travels up and down. There are two valves, an inlet and an exhaust valve, and an arrangement to open and close them. The idealized cycle is illustrated in Figure 1.2.

In Figure 1.2 we plot the pressure in the cylinder against the volume in the cylinder. Notice that the piston does not go quite all the way to the top of the cylinder; the piston is at the top of its travel at 0, 2 and 3, and there is a small space still above it, the combustion chamber. At the beginning, the piston is at the top of its travel. This is known as Top Center (formerly Top Dead Center) or TC (formerly TDC). Initially, the inlet valve is open and the exhaust valve is closed. In the idealized cycle, the crank rotates and the piston moves from the top of its travel to the bottom, corresponding to travel from 0 to 1 on Figure 1.2, drawing in a charge of air at atmospheric pressure. The position at the bottom is known as Bottom Center (formerly Bottom Dead Center) or BC (formerly BDC). This stroke from TC to BC is known as the intake stroke. When the piston reaches the bottom, at 1 on Figure 1.2, the inlet valve closes and the crank continues to rotate, returning the piston to the top of its travel and compressing the air. This is known as the compression stroke. The closing of the valve is another idealization – valves cannot close instantly (we will return to that later). This corresponds to the travel from 1 to 2 on Figure 1.2. The air is heated by the compression, but the compression is imagined to be without heat loss to the cylinder walls, or adiabatic. When the piston reaches 2 we imagine adding a certain amount of heat all at once. This heat corresponds to the heat that would have

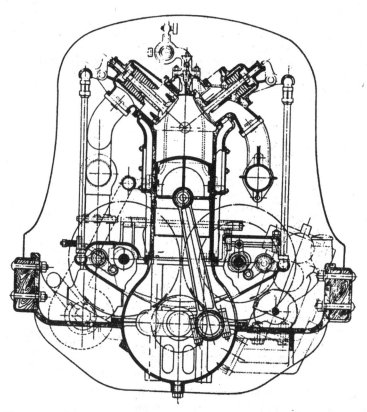

Figure 1.1. Cross-section of the Belgian Pipe engine, which appeared in 1905. It has a hemispherical combustion chamber and overhead valves operated by pushrods. This is the first example of this configuration [17]. Reproduced by kind permission of the Estate of Griffith Borgeson.

been released by the burning of the gasoline that would have been drawn into the cylinder with the air in a real engine. However, here we have replaced that process (which we will discover is extremely messy) by some imaginary process involving no gasoline and all the heat is released at once. This is an idealization also, because in a real engine the charge is burned gradually. This heat addition corresponds on Figure 1.2 to moving from 2 to 3. Still, we are imagining that there is no heat loss to the walls of the cylinder. The crank continues to turn and the piston travels down again, the heated gases expanding, moving from 3 to 4 on Figure 1.2. This is known as the power stroke. At this point the exhaust valve opens (again, instantly), and the pressure in the cylinder drops instantly to atmospheric pressure, moving from 4 to 1 on Figure 1.2. Finally, the crank continuing to turn, the piston rises to the top again, to 0, expelling the remaining gases through the open exhaust valve. This is the exhaust stroke.

We have suggested here some of the ways in which the real cycle probably differs from the ideal cycle, and we will go into that in much greater detail later. However, we can mention some of the other engine types which have ideal cycles that differ from the Otto cycle described here. The Diesel cycle offers a good example. To understand the Diesel cycle, we must define the compression ratio. Compression ratio compares the volume of the cylinder when the piston is at the bottom of its travel (BC) to the volume of the combustion chamber, or the volume of the cylinder when the piston is at the top of its travel (TC). The temperature of the compressed gases at 2 in Figure 1.2 is obviously a function of the compression ratio. In the Diesel cycle the compression ratio is much higher

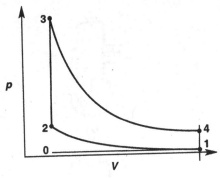

Figure 1.2.   The Ideal Otto cycle.

than in the Otto cycle, so the temperature of the compressed air is very high. The fuel in a real diesel engine is sprayed into the combustion chamber; when the fuel spray meets the high-temperature air, the spray droplets begin to evaporate and the vapor mixes with the air. Because the air temperature and pressure are above the fuel's ignition point, after a brief delay of a few crank angle degrees the vapor-air mixture bursts into flame. The spray continues for a while as the piston descends. The idealization of this cycle has a horizontal line from 2 to 3, the heat being added at constant pressure rather than constant volume as in the Otto cycle.

In one type of Hesselman engine, the fuel is sprayed directly into the cylinder, or rather into a small pre-combustion chamber connected to the cylinder by a small orifice. The compression ratio is not high enough to generate temperatures that will ignite the fuel spray, so a spark is used. Again, the spray can continue for a while as the piston descends. The pressure in the pre-combustion chamber is released more or less slowly into the cylinder through the orifice, so that the cylinder pressure rise is neither as abrupt as the Otto cycle nor as flat as the diesel cycle, but somewhere in-between. There is no simple idealized cycle that models this engine. The Hesselman engine was widely used for farm tractors and had the advantage that it would run on nearly any fuel for reasons we will see later. The Hesselman engine was the first of what are now known as Gasoline Direct Injection engines, or GDI, which we will discuss in Chapter 5, Section 5.12.

Many modern engines have pre-combustion chambers, and some use fuel injection directly into the cylinder, so that the pressure rise in these engines is somewhat modified in the direction of the Hesselman engine. Hence, quite aside from the idealizations mentioned above (which involve heat transfer and valve opening among others) real modern engines do not fit the Otto cycle very well.

The ideal Otto cycle that we have described is a four-stroke cycle because the piston makes a total of four strokes — two up and two down, including the intake and exhaust strokes. There is another spark-ignition engine with virtually the same ideal cycle. This is the so-called two-stroke cycle. In this engine, there are no intake and exhaust strokes; the intake valve is replaced by ports in the cylinder wall, which are uncovered by the descending piston. The exhaust

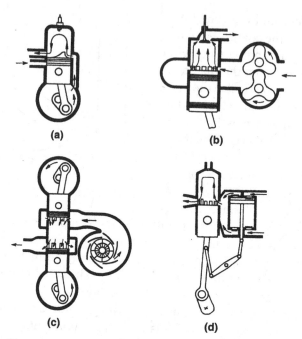

Figure 1.3. Various arrangements of two-stroke engines [94].
Copyright © 1968 and new material © 1985 by The Massachusetts Institute of Technology.

valve may also be replaced by ports. The engine may have (see Figure 1.3) a scavenging pump, or one of the other arrangements of Figure 1.3 may be used. The arrangement (b) was used in the General Motors diesel engines which were manufactured in three, four, five and six-cylinder models for use in buses and trucks. We will come across these again, because the Roots blowers used with these engines generally survived their parent engines and were reused on fuel dragsters.[1]

Whichever system is used, it maintains a positive pressure in the intake manifold so that when the intake port is uncovered (the exhaust valve being open) the fresh charge blows through the cylinder, blowing the exhaust gases out the exhaust valve. Presuming that the scavenging system does not generate any significant pressure, the ideal cycle for these engines is very similar to the Otto cycle (Fig. 1.2), without the leg between 0 and 1.

The two-stroke cycle engine is the one so familiar from lawn mowers and chain saws. This is type (a) from Figure 1.3, where the scavenging is provided by compression of the intake charge in the crankcase. This is cheap, but has the disadvantage that normal lubrication is not possible, because the oil would be picked up and carried into the cylinder by the incoming charge. Instead, the

---

[1] A fuel dragster has a highly modified engine, usually supercharged, running on exotic fuels, with tractor wheels at the rear and bicycle wheels at the very distant front, and designed solely to compete in the standing quarter-mile.

gasoline is mixed with oil, some of which stays in the crankcase when the charge goes through and lubricates the bearings. Saab had a very successful three-cylinder two-stroke cycle engine, crankcase-scavenged, which was widely used in racing; we may conclude that properly tended to, the lubrication system works well. I am sorry to admit that I and my family were responsible for the destruction of two of these engines due to inattention to adding oil to the gasoline.

These two-stroke engines have always had a reputation for polluting the atmosphere due to the oily exhaust. The gas flow, particularly of type (a) in Figure 1.3, is quite complex, and there is considerable mixing of the incoming charge with the outgoing exhaust. Consequently, these engines have not been able to meet the U.S. federal air pollution standards. Type (b), however, is considerably better, and recently Daewoo has announced a two stroke engine of type (b) which does meet the U.S. air pollution standards. Two-stroke engines have about 1.5 times the power of four-stroke engines of similar displacement, because they have twice as many power strokes per unit of time, but work must be done to pump the fresh charge into the cylinder.

1.2
EFFICIENCIES

1.2.1 Air Cycle Efficiency

Because the pressure is higher on the power stroke (due to the addition of heat at TC) than it was on the compression stroke there is net work done by the cylinder during the cycle. The work is equal to the area within the diagram of Figure 1.2. At the same time, not all the heat that was added at TC has been converted to work, because the air in the cylinder at 4 on Figure 1.2 is still hotter than it was at 1. When the exhaust valve is opened at 4, this heat is allowed to escape. Students of thermodynamics know that this is inevitable–some heat must always be rejected.

We can define an efficiency for this process based on this idealized cycle (known as the air cycle, since the working fluid is presumed to be air all the way through the cycle). This is a relatively straightforward calculation (which can be found in [94]): an isentropic compression from 1 to 2, and an isentropic expansion from 3 to 4. The efficiency is designated by η_{ac}, where the subscript refers to the air cycle, and it is equal to the work produced, divided by the heat added. The result of the calculation is:

$$\eta_{ac} = 1 - \left(\frac{1}{r}\right)^{k-1} \tag{1.1}$$

where r is the compression ratio, and k is the ratio of specific heats, $k = c_p/c_v$, the specific heats at constant pressure and at constant volume. In a diatomic gas at normal temperatures $k = 1.4$. The air cycle efficiency ranges from 0.42 at a compression ratio of 4 to 0.56 at a compression ratio of 8.

1.2.2 Real Gas Efficiency

Now, in fact, η_{ac} is not much like the actual efficiency of the real process. As we pointed out previously, many things are happening in the real engine that are not included in η_{ac}. In the first place, the gas is not air. On the intake and compression strokes air is mixed with fuel, and on the power and exhaust strokes the gas is a mixture of C, CO, CO_2, H, OH, H_2, H_2O, N, NO, NO_2, N_2, O, and O_2. This makes a substantial difference; for example, a stoichiometric gasoline mixture has a ratio of specific heats $c_p/c_v = 1.35$ [94], rather than 1.4. In addition, there is heat loss taking place and the combustion does not occur instantly.

We can calculate an efficiency taking account of the real gases, but ignoring the heat loss and the fact that the combustion is not instantaneous. That is, the heat is liberated by combustion of the mixture at constant volume. The walls are taken to be adiabatic. This efficiency takes account of some of the complexities of the real situation, but not all of them. Unfortunately, this is a much more complicated calculation and does not result in a simple formula like Equation (1.1). Among other things, the equilibrium composition of the mixture of gases during the power and exhaust strokes is a function of temperature and is constantly changing. This efficiency is called the real gas thermodynamic efficiency, and is designated by η_o (the subscript o for Otto). The ratio of η_o to η_{ac} is only a very weak function of compression ratio; it is primarily a function of ϕ (the fuel:air ratio relative to stoichiometric), because this changes the value of c_p/c_v, and is essentially independent of inlet and exhaust conditions over the range of values normally encountered. Very roughly (from experimental data, mostly on Coordinating Fuel Research engines), we may write for $\phi = 1.0$ [94]

$$\frac{\eta_o}{\eta_{ac}} = 0.69 \tag{1.2}$$

1.2.3 Indicated Efficiency

Finally, we have the effects of heat loss, the fact that the combustion does not occur instantly, and so forth, which we sketched out in Section 1.1. The actual efficiency including all these effects is called η_i, where the subscript refers to *indicated*. The word indicated means obtained from an *indicator diagram*, which is produced by a recording device attached to the cylinder that registers actual pressures and volumes within the cylinder. Hence indicated means actual, real, but measured at the cylinder, as opposed to at the flywheel. That is, it does not contain the mechanical efficiency: the energy expended in pumping the gases into and out of the cylinder, driving some of the accesories (always the oil pump, usually not the water pump, fan and generator), turning the camshaft (to operate the valves), rubbing the piston rings up and down against the cylinder walls, and turning the crankshaft in its bearings. Determining η_i is extremely complex and must be obtained from experimental data. However, as was the case with η_o, its ratio to the other efficiencies does not vary much; it is primarily a function of ϕ and weakly of combustion chamber shape, which influences heat loss through the surface:volume ratio. If we take data from a CFR (Coordinating Fuel Research) engine, which has a variable compression ratio and a cylindrical combustion chamber, and has the valves in the head, we obtain for $\phi = 1.13$ [94]

$$\frac{\eta_i}{\eta_o} = 0.86 \tag{1.3}$$

[94] This means that the ratio we are interested in is approximately

$$\frac{\eta_i}{\eta_{ac}} = 0.69 \times 0.86 = 0.59 \tag{1.4}$$

In Equation (1.18) we will give an interpolation formula for the effect of mixture strength on this ratio.

Using the ESP (see the Preface, and Chapter 8), and simulating a Jaguar XK engine with hemispherical combustion chamber at stoichiometric conditions (we will discuss this in greater detail later), we find a value of $\eta_i/\eta_{ac} = 0.603$, while our interpolation formulas Equation (1.4) and Equation (1.18) suggest a value of 0.594. As you can see, there is not much difference. The slightly higher (2%) value of the ratio is probably not significant, considering the effects that have been omitted from ESP, although it is of the right magnitude and in the correct direction:the XK engine does have a slightly more efficient combustion chamber shape, with a smaller surface:volume ratio.

1.3
A MORE REALISTIC CYCLE

Now let's talk in detail about the ways in which reality differs from the ideal cycle, the phenomena that are responsible for η_i differing from η_o. In Figure 1.4 we have sketched a cartoon of a real cycle superimposed on an ideal cycle, to show the differences more clearly. In a little while we will show something more realistic than this cartoon. The curve labeled y-z is an isentropic curve through the point b. We have labeled the points where ignition occurs (a), the end of combustion (b), and the opening of the exhaust valve (c). The hatched area to the left of the curve a-b and below the curve y-b is called *time loss*, the hatched area above the curve b-c is called *heat loss* and the stippled area to the right of c-1 is called *exhaust blowdown loss*. We will discuss these below. We have not indicated on Figure 1.4 the ways in which the real intake and exhaust strokes (0-1 on the Figure) differ from

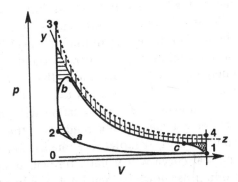

Figure 1.4. Cartoon of real Otto cycle superimposed on ideal. [94]. Copyright © 1968 and new material © 1985 by The Massachusetts Institute of Technology.

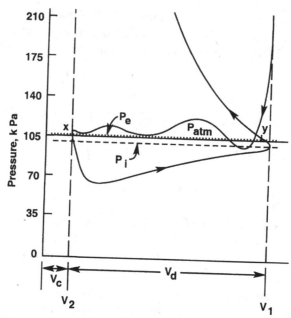

Figure 1.5. Inlet/exhaust part of the indicator diagram at partial throttle [94]. Copyright © 1968 and new material © 1985 by The Massachusetts Institute of Technology.

the idealization. These are normally included in the Mechanical Efficiency (see Figure 1.5).

1.3.1 Time Loss

Combustion is not an explosion. It does not occur at once. It is, in fact, an orderly burning of the mixture of gasoline vapor and air. The spark ignites the mixture in the immediate vicinity of the spark plug, and a flame front proceeds spherically outward from there. The flame front propagates at a more-or-less uniform speed determined primarily by the turbulent velocities in the gas. Ahead of the flame front is unburned mixture, and behind it are the products of combustion. It takes the flame front a certain time to travel from the spark plug to the farthest reach of the combustion chamber. As the flame front travels it is converting chemical energy to sensible heat, and so the temperature and pressure in the gas are rising continuously. It is necessary to start the combustion (ignition) considerably before TC, and even so the combustion will go on substantially past TC. The greatest efficiency usually is obtained when the point of ignition and the point at which combustion is complete are roughly symmetric with respect to TC. In any event, the real curve will usually lie well within the ideal curve (except at very late timing), as shown in Figure 1.4, and hence the difference in area represents work that cannot be extracted. This is known as time loss, meaning that it is a loss due to the finite time that the flame front takes to cross the combustion chamber. Of the difference between η_i and η_o, roughly 30% is time loss [94]. The definition is a little misleading because the mixture is losing heat during the combustion so that the point b is well below the curve 3-4 (if combustion were considered with finite burn time, but no heat loss, the point b would be above the line 3-4). How to separate the effects of heat loss from those of finite burn time in an unambiguous way is not clear. That is, just how much of the area between b and 3 to apportion to time loss and how much to heat loss is not settled. We will suggest a workable possibility below.

1.3.2   Heat Loss

As the intake charge is compressed on the compression stroke its temperature is rising, and it is consequently losing heat to the cylinder walls. This is not too serious, because its temperature is not yet very high; the reduction of the pressure and temperature at the ignition point is negligible. After the combustion, however, its temperature is considerable, and as it expands on the power stroke, its temperature dropping, it is losing considerable heat to the walls of the cylinder and combustion chamber, resulting in a substantial reduction in its temperature and pressure at the end of the stroke. (We have seen also that it lost considerable heat during the combustion.) The difference between this and the ideal cycle represents work that cannot be extracted. The difference between the real and ideal cycle on the compression stroke is work that need not be done on the gas, but this is much smaller than the difference on the power stroke and is usually negligible. The combination of the two is known as heat loss. Of the difference between η_i and η_o, roughly 60% is heat loss [94]. The usual way [94] to separate time loss from heat loss is shown in Figure 1.4: the horizontally hatched area is given to time loss, and the vertically hatched to heat loss, where the line y-z is an isentrope through the point b. This is referred to as *apparent* time loss and *apparent* heat loss [94].

1.3.3   Exhaust Blowdown Loss

As the piston approaches BC on the power stroke the exhaust valve is opened. Typically, on a street car the exhaust valve opens some 47° before BC. Immediately the pressure begins to drop as the exhaust rushes out of the cylinder. This difference in pressure between the real and ideal cycle at this point in the cycle represents unavailable work, and is called exhaust blowdown loss. Of the difference between η_i and η_o, 10% is exhaust blowdown loss [94].

1.3.4   Other Losses

Adding together time loss, heat loss and exhaust blowdown loss, it appears that we have accounted for one-hundred percent. Well, in fact, not quite. The figures given above are approximate, and there are other losses, usually quite negligible at full throttle but definitely not at partial throttle. Full throttle is not a typical operating point; most driving is done at high manifold vacuum and this changes things considerably. Notice in Figure 1.5 (at partial throttle) that the pressure drops below atmospheric as the piston descends on the intake stroke, drawing air into the cylinder, and it is above atmospheric on the exhaust stroke in order to expel the gases from the cylinder. This loop, consisting of exhaust and intake, represents work that must be done on the gases, and hence is lost energy. These losses are called pumping losses. In Figure 1.5 we show this part of an indicator diagram, greatly expanded. Under these conditions the peak pressure on the exhaust stroke is of the order of 120 kPa while the minimum pressure on the intake stroke is of the order of 70 kPa, giving a peak pressure difference on the order of 50 kPa. At WOT (Wide Open Throttle) the exhaust pressure is a little higher because the mass flow is greater, while the minimum pressure on the intake stroke is considerably higher (the manifold vacuum is quite low) so that the peak pressure difference drops to perhaps 40 kPa [47].

In Figure 1.6 we show an indicator diagram for a typical engine at WOT. It is clear that the peak pressure rises to something on the order of 5.2 MPa at wide open throttle, giving a difference from atmospheric of 5.1 MPa. Hence, at WOT the peak pressure difference on the intake and exhaust strokes is on the order of 0.4% of the peak combustion pressure. It is clear from Figure 1.6 that when the intake and exhaust portion of the cycle are shown on the same plot, this part of the curve appears practically flat. Hence, the work represented by the area in this part of the curve is a small fraction of the total, no more than a few percent at WOT. On the other hand, at idle the peak intake-exhaust pressure difference can rise as high as 80 kPa, while the peak combustion pressure over atmospheric can be as low as perhaps 135 kPa (because so little fuel is being burned), so that the ratio is nearly 60%. Hence, at anything but WOT, the work done on the intake and exhaust strokes is significant.

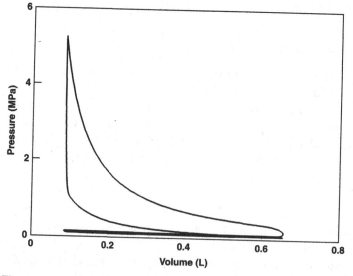

Figure 1.6. Indicator diagram for a Jaguar XK engine (3.4 L, $r = 8$, paper element air cleaner), generated by ESP.

By convention this pumping work is included in the friction and ends as part of the mechanical efficiency. If the average cylinder pressures are known or can be estimated during the intake and exhaust strokes, the pumping work can easily be estimated, as the pressure difference times the displacement. In the case of a diagram like Figure 1.5, if we take the average as one-half of the peak, and suppose the peak is 40 kPa, corresponding to WOT, we have an average pressure difference of 20 kPa. Let us suppose that the diagram of Figure 1.5 was obtained from the engine of Figure 1.6. The displacement of this cylinder is 5.74×10^{-4} m^3, so that the work done is 11.5 J. On the other hand, ESP tells us that the net work output is 480 J. The pumping losses are thus about 2.4% in this case. The value generated by ESP for Figure 1.6 (corresponding to quite short intake and exhaust manifolds – that is, no manifold tuning) gives about 2%.

The pumping losses become a substantial fraction of the total at partial throttle, when manifold vacuum is high. Then the size of the main part of the cycle, 1-2-3-4 on Figure 1.2, is greatly reduced. In fact, at idle the engine produces no net work, and the work produced in the cylinder must just be equal to the mechanical losses (pumping, accesories, friction). Hence, the area of the 1-2-3-4 part of the curve, less the area of the intake – exhaust part, must be equal to the accessories and friction. In this case, the intake – exhaust part is certainly not negligible.

There are two other losses, both of which are small under normal circumstances. The first loss is because not all the fuel that is brought into the cylinder is burned, for several reasons. Mixing of the gasoline vapor and the air is very non-uniform and turbulent, and some regions are lean and some are rich. Also, in a carburetted engine, some cylinders are lean and some are rich – the mixture is adjusted so that the leanest burns satisfactorily. Because of poor mixture control, either within a cylinder or from cylinder to cylinder, some of the mixture is not burned. Finally, some of the charge next to the cylinder walls and in crevices around the spark plug and the valves is chilled, and will not burn. For all these reasons, a small fraction of the fuel brought into the cylinder is not burned and leaves with the exhaust. These are the unburned hydrocarbons that make such a major contribution to air pollution, but they also decrease efficiency because they represent energy that is not turned into work. Combustion efficiency is defined as the ratio of the energy released by burning to the energy that was brought into the cylinder. For mixtures that are stoichiometric or somewhat leaner (but not so lean as to misfire), this averages about 0.98 for most engines (although it has been observed to be as low as 0.9). As the mixture becomes richer, the amount of fuel that can be burned is limited by the air available. If all the air were used the combustion efficiency would be $\eta_c = 1/\phi$, where ϕ is the equivalence ratio, the ratio of the fuel:air ratio to stoichiometric. In fact, η_c falls a little faster than this, so that even the percentage of this figure decreases. By $\phi = 1.2$, η_c is down to 0.777 on the average (with a spread of approximately ±6%). Hence, only 93% of the fuel theoretically burnable is being burned [47] (i.e., $0.93 \times 0.83 = 0.77$).

Finally, we have leakage losses. The piston rings, and sometimes the valves, do not seal perfectly and consequently the cylinder pressures do not rise as high as they should, representing a small loss. For an engine in good condition this is negligible, but for an old, worn engine, blowby (as it is called – the loss past the rings) results in a significant loss of power. Because it takes a finite time for fluid at an elevated pressure to leak through an orifice, the leakage losses are usually less serious at higher engine speed; an old, worn engine runs better at higher speed. In the same way, many racing engines use only one piston ring in order to reduce the friction at higher piston speeds, the resulting leakage not being a problem at operating speeds.

1.4
KNOCKING

As the flame front travels across the combustion chamber the temperature and pressure of the gases ahead are rising, due to the continual release of heat. Above a minimum value of temperature and pressure, a mixture of fuel vapor and air with a particular concentration at a particular pressure and temperature has a delay time after which it will spontaneously ignite, known as autoignition. That is, when the mixture reaches the particular temperature and pressure, a reaction begins, the end point of which is ignition. This reaction does not take place all at once. This phenomenon is a function of pressure and concentration, but that need not concern us here. If the unburned mixture in front of the advancing flame (known as the end-gas) in the combustion chamber were to reach its autoignition point simultaneously due to the temperature and pressure rise, then the entire remaining unburned mixture would ignite at once, not waiting for the arrival of the flame front. The combustion which previously had been an orderly progressive burning would now become an explosion. In fact, it does not all ignite simultaneously because the composition and the temperature are not quite uniform due to turbulent mixing; as a consequence, one or more regions will reach their autoignition point slightly before the others and will ignite first. Because, however, all the end gas was very close to its autoignition point other regions will soon reach their autoignition points and the entire end gas will ignite, if not instantly, then in a time shorter than the time in which the flame front normally crosses the combustion chamber. This can be seen by the more rapid pressure rise in part (a) of Figure 1.7. This vision of knocking is supported by numerous experimental studies making use of Schlieren, shadowgraph and optical fiber techniques [60], [72], [73], [87], [90], [91], [92].

We also show a series of photographs of flame propagation in a knocking engine cylinder.

In Figure 1.8 the spot which first ignites in the end gas is marked with an x – it is clear that the remainder of the unburned mixture ignites quite rapidly. If this explosion can be clearly heard, it is variously called a knock, a ping, or in England, a pink. Knocking was formerly called detonation, but that is now reserved for steady-state combustion situations in which the flame front propagates at supersonic speeds (when the flame front propagates at subsonic speeds, it is called deflagration). In autoignition of the end gas, what is important is the unsteady

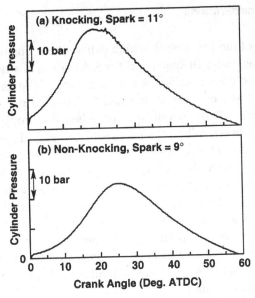

Figure 1.7. Examples of knocking and non-knocking pressure traces (1500 rpm, 12.5 A/F ratio, wide open throttle, two-valve cylinder head. From [87]. Reprinted with permission from 960497 © 1996 Society of Automotive Engineers, Inc.

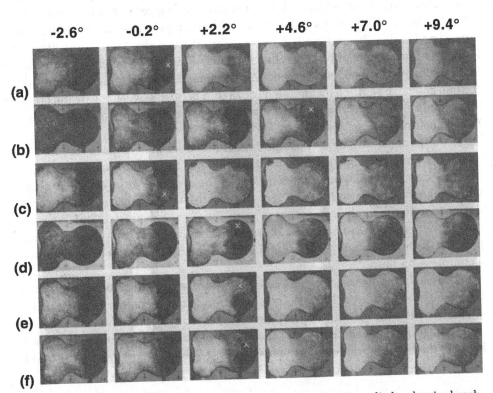

Figure 1.8. Six series of photographs of the flame in a glass-topped engine cylinder, showing knocking. The point where the autoignition first appears is indicated by x in each series. Angles are crank angle degrees after top center [106]. Reprinted with permission from 360126 © 1936 Society of Automotive Engineers, Inc.

and chaotic nature of the process, in which different regions trigger autoignition of other regions chaotically in space and time. Knock is always due to autoignition, but autoignition does not always produce knock. The flame front from the autoignition sites in the end gas is sometimes found to propagate at subsonic speeds and sometimes at supersonic speeds. When the propagation is subsonic the knocking is less severe. When the propagation is slightly supersonic (known as developing detonation) the knocking is more severe, and results in fluctuations in cylinder pressure that can be seen in part (a) of Figure 1.7. As the propagation is more and more supersonic, the knocking becomes more and more severe.

What is heard during knocking is the structural vibration of the engine block, excited by the fluctuations in cylinder pressure. Knock sensors,[2] which are used to control knocking (by retarding the spark), operate by sensing this vibration in the engine structure. Severe knocking can be very damaging, particularly to the pistons, causing damage to the top land (the land is the high part between grooves; the top land is the part of the piston between the top groove and the top of the piston) and ring groove [87], and (after long periods) surface damage to the crown leading to erosion and ultimately failure of the piston crown [94].

It is clearly desirable to avoid knocking. For a given fuel with given anti-knock properties (that is, a given value of autoignition time delay at cylinder pressures, temperatures and concentrations), it is essential to control either the peak pressures in the cylinder, because these control the temperatures, or to control the burn time (so that the end gas is burned before it has a chance to autoignite), or both. The burn time can be controlled by controlling the turbulence level, combustion chamber shape and number and placement of spark plugs. In addition, the end gas can be chilled by forcing it into a thin region between the piston crown and cylinder head; chilling slows the autoignition reaction so that the end gas is burned before it can autoignite.

In designing an engine, the compression ratio is selected to make sure that the peak pressure is not high enough to cause knocking, taking into consideration the other factors. For example, immediately after the Second World War in England, when the available gasoline had very poor anti-knock performance, cars were designed with relatively low compression ratios. The Lagonda 2.6 Liter (designed during the war for manufacture in 1947), for example, had a compression ratio of only 6.5. Bear in mind, however, that this is not the only factor in the selection of a compression ratio.

For an existing engine, the spark advance is the single factor having the greatest influence over knocking. If the point of ignition is too far before TC, that is, too advanced, then the combustion will be complete roughly at TC and this will be the point of peak pressure. That is the worst situation for knocking, because it leads to the highest pressure and temperature. As the spark (point of ignition) is

---

[2] Knock sensors were used first in super/turbocharged engines to control knock due to the much higher cylinder pressures generated, then in premium-fuel engines to control knock in case they were run on regular fuel. Today they are used in nearly all engines, sometimes one per bank in V engines.

retarded (that is, as the point of ignition is moved closer to TC) combustion is completed after TC, and because the piston has already started down, the peak pressure (and temperature) are not as high. As the spark is retarded more and more the peak pressure occurs later and later and becomes lower and lower. At one time, with a given engine and a given fuel, the distributor was adjusted by ear to an advance that was as great as possible without producing knock. Now, things are more complicated, and there are many other considerations involving the catalyst, emissions, warmup, and so forth. Engine control computers in engines with knock sensors make these calculations.

In 1.9 we show several indicator cards generated by ESP for a CFR engine (bore of $b = .0825$ m, stroke of $S = 0.114$ m, with a compression ratio of $r = 6.0$. Throughout, we will use b for bore and S for stroke.). The spark advance is set respectively at $0°$, $18°$, $26°$, $39°$ BTC (before top center). The number of crank degrees occupied by the combustion is approximately the same in all cases, but it is evident how much the peak pressure is affected.

1.5
MEAN EFFECTIVE PRESSURES

It is helpful to define several quantities that we will call mean effective pressures. A *mean effective pressure* is the constant pressure which, acting on the piston area through the stroke, would produce the observed work per cycle. This pressure is a fiction, but a useful one. During a real cycle the pressure is only significant during the compression and power strokes. During the compression stroke work is done on the gas in the cylinder, while during the power stroke the gas in the cylinder does work on the piston. Hence, the net pressure would be the difference between these two. We would, therefore, expect the mean effective pressure to be about the difference between the average pressures on the power and compression strokes. Formally, we define

$$P = mep\, A_p\, Sn\frac{N}{X} = mep\, V_d\frac{N}{X} \tag{1.5}$$

where P is the power, mep is the mean effective pressure, A_p is the piston area, S is the stroke, n is the number of cylinders, N is the rotational speed of the crankshaft (rotations per unit of time) and X is the number of revolutions per power stroke. The quantity $X = 2$ for a four-stroke cycle engine, and $X = 1$ for a two-stroke cycle engine. V_d is the total displacement of the engine. Mean effective pressures are measured at various places.

1.5.1 A Word On Units

We must face the fact that the automobile, being old technology, carries a lot of old baggage. One piece of old baggage is the use of outmoded units. Although the U.S. Society of Automotive Engineers has resolutely adopted SI units, so that everything is measured in watts, kilograms, meters, and so forth, virtually everybody else who is concerned with cars (magazine writers, enthusiasts, racing buffs) still speaks in horsepower, horsepower per square inch, horsepower per liter(!), piston speeds in feet per minute, *bmep* in psi, and so forth. This makes calculation a nightmare because everything must first be translated into a consistent system of units (say, SI) and the calculation carried out, and the answer then translated into the local dialect. The SAE presumably hopes that persistent use of SI units will eventually make *bmep* in Pascals and specific output in kW/cm^2 comprehensible to the natives. I hope so, though I will mourn the passing of the older units as I will that of Gaelic and Basque. In this book I will use SI units, but I will give translations because some numerical values in the older units were historical high water marks and still have great evocative power.

1.5.2 Brake Mean Effective Pressure

The most common mean effective pressure is the Brake Mean Effective Pressure, or *bmep*. The adjective brake refers to measurement at the flywheel. Originally, power output was measured by applying an ordinary brake to the flywheel, the brake being attached to a long arm, and the moment produced was measured. The power produced by the engine is proportional to the product of the moment produced and the angular rotational speed. This brake was a primitive dynamometer, and was developed by François Marie Riche, Baron de Prony (1755–1839), a French mathematician and engineer; it was known as a Prony brake. At present brake power simply means power including the effects of mechanical inefficiency – that is, loss of power to pumping, valve gear, certain accessories and friction. This is usually designated P_b (though sometimes HP_b for brake horsepower).

A typical WOT *bmep* at present lies between 0.9 and 1.1 MPa (between 140 and 160 psi). Average values of *bmep* have been rising fairly steadily during the development of the automobile. U.S. passenger cars in 1950 had values averaging around 0.75 MPa (110 psi), ranging from 0.65 to 0.9 MPa (95 psi to 130 psi). The *bmep* rises with compression ratio; however, as we shall see later, there are a number of other reasons for the gradual rise of *bmep*.

A physicist examining a problem looks for dimensionless groups to characterize it. Unfortunately, there do not seem to be any very useful dimensionless groups to characterize the spark-ignition engine. However, *bmep* is useful because it is roughly comparable even in very different engines, as these different engines burn the same fuel, necessarily under approximately the same conditions, and hence generate similar pressures.

Differences in *bmep* represent genuine design differences, and not irrelevant differences such as size. Thus, we can compare [94] a tiny model airplane engine of displacement 1.6 cm^3 (0.1 cubic inch) running at 11,400 rpm, with a huge stationary diesel engine with a displacement of 0.4 m^3 (26,500 cubic inches) running at 164 rpm. The first has a *bmep* of 0.32 MPa (47 psi), while the second has a *bmep* of 0.45 MPa (66 psi).

The relatively small difference between these two figures is probably a function of real design differences between the two engines (some of which are still indirectly size-related), possibly the compression ratio. However, [94] mentions a number of respects in which very small engines are at a disadvantage, notably a large surface:volume ratio in the combustion chamber and cylinder (proportional to the inverse of a characteristic length), which will result in a much higher heat loss and consequent reduction of the *bmep*. In addition, Reynolds numbers of the gas flow in a small engine are low and the time for mixing is short, resulting in very poor mixing and vaporization of the fuel.

1.5.3 Indicated Mean Effective Pressure

A mean effective pressure can also be defined in the cylinder, rather than at the flywheel. This is called the indicated mean effective pressure, or *imep*. The *imep* clearly does not contain the effects of mechanical friction, or the work necessary to pump the gases in and out and drive the camshaft and various accessories. It consequently will be somewhat higher than the *bmep*. The defining equation is the same, but the P_i, the indicated power, rather than the P_b, the brake power, is used.

Later, when we discuss friction, we will define a friction mean effective pressure, or *fmep*, which is defined using the P_f or the power expended on friction. Note that

$$bmep = imep - fmep \tag{1.6}$$

That is: the brake work (what you actually get) is the difference between the indicated work (what you think you should get) and the frictional losses.

1.6
PISTON SPEED

Another variable that is more-or-less comparable among engines of otherwise very different natures is the average piston speed. Piston speed is proportional to engine speed and is averaged over a cycle. In the course of this book we will see a number of reasons why average piston speed is comparable among different engines. We will preview them here. Probably the most important has to do with the gas flow through the valves. We will find that a fundamental limitation on the maximum gas flow is the occurrence of sonic flow in the valve aperture, and we will be able to relate this to the average piston speed. In addition, we will find that the mechanical efficiency, normally in the neighborhood of 0.85 for low engine speeds, drops dramatically at high piston speeds, reaching 0.6 in the neighborhood of an average piston speed of 20 m/s (4000 ft/min). 20 m/s represents a practical maximum for most engines. This is independent of engine size because engines of all sizes have piston rings made of roughly the same material, rubbing against cylinder walls of roughly the same material, bearings of similar design using oil of similar viscosity, and so forth. When we discuss heat transfer we will see that the temperature rise of parts such as exhaust valves and piston crowns is proportional to the turbulent velocity, which is roughly proportional to the piston speed in a given engine. Finally, the inertial stresses in connecting rods (and other reciprocating parts) are proportional to the square of the piston speed. For all these reasons (and other secondary ones, including ring float and valve float), piston speed is much more nearly comparable (than other variables) among engines of otherwise very different characteristics. For example, the tiny airplane engine has a mean piston speed of 4.9 m/s (980 ft/min), while the enormous stationary diesel has a mean piston speed of 5.5 m/s (1100 ft/min).

The mean piston speed is easy to calculate. If N is the rotational speed, say in revolutions per second, or rps, then $1/N$ is the time for one revolution. Half of this, $1/2N$ is the time for one stroke. If the stroke length is S, then the average speed $\bar{V}_p$ is

$$\bar{V}_p = \frac{S}{1/2N} = 2NS \tag{1.7}$$

If we examine our expression for *bmep*, Equation (1.5), we see that it can be rewritten in terms of the mean piston speed, to give

$$P_b = bmep\, A_p n \frac{\bar{V}_p}{2X} \tag{1.8}$$

1.7

SPECIFIC POWER

Now let us examine Equation (1.8). If *bmep* removes irrelevant differences among engines (such as physical size), and reflects only genuine design differences, and if the same can be said of $\bar{V}_p$, then the quantity $P_b/A_p n$, the power produced per unit piston area must also be such a quantity. It is called the specific power.

$$\frac{P_b}{A_p n} = bmep\, \frac{\bar{V}_p}{2X} \tag{1.9}$$

In fact, the three quantities: *bmep*, $P_b/A_p n$ and $\bar{V}_p$ are not independent, any two being sufficient to determine the third through Equation 1.9 (with the addition of the information that the engine is four-stroke or two-stroke). Note that, for similar $P_b/A_p n$ and $\bar{V}_p$, the *bmep* for a four-stroke engine will be twice the *bmep* for a two-stroke engine.

We measure $P_b/A_p n$ in kW/cm^2, and *bmep* in MPa. In the older literature the quantity $P_b/A_p n$ is always presented in HP/in^2, and *bmep* is presented in psi. (1.00 HP/in^2 = 0.116 kW/cm^2.) As we have already seen, typical values of *bmep* are about 1.0 MPa = 150 psi. A typical plot is *bmep* versus $\bar{V}_p$, with lines of constant $P_b/A_p n$. We give here one such plot, taken from [94], with some additions. Notice that the 1954 passenger cars have *bmep* clustered around 0.76 MPa (110 psi), specific outputs clustered around perhaps 0.24 kW/cm^2 (2.1 HP/in^2) and piston speeds of perhaps 13 m/s (2600 ft/min). The 1997 performance cars for the most part have *bmep* clustered around 1.1 MPa (160 psi), piston speeds clustered around 15 m/s (3000 ft/min) and specific outputs in the neighborhood of 0.38 kW/cm^2 (3.3 HP/in^2). The only exceptions are the Ford Mustang (3), which finds itself nearly among the 1954 passenger cars, and the Saab turbo (11), competing with the transport aircraft. All but the Saab are normally aspirated. The Ford Mustang (3) and the Pontiac Grand Prix (9) have two valves per cylinder; all the others have four valves per cylinder.

1.8
STROKE/BORE RATIO

I have included in the caption of Figure 1.9 the power per unit displacement of the 1997 cars in kW/L (HP/L). This is another traditional measure of output, though not as fundamental as – and not to be confused with – specific output (power per unit piston area), because the power per unit displacement depends on the length of the stroke. If we take Equation 1.8 and divide by the displacement $V_d = A_p n S$ we have

$$\frac{P_b}{V_d} = \frac{bmep \, \overline{V}_p}{S2X} \tag{1.10}$$

We can see from Equation 1.10 that other things being equal (that is, for the same *bmep* and $\overline{V}_p$, which we expect to be comparable among engines), an engine with a shorter stroke will have a higher power per unit displacement. There has been a progressive reduction in the stroke:bore ratio, S/b, over the decades. Among

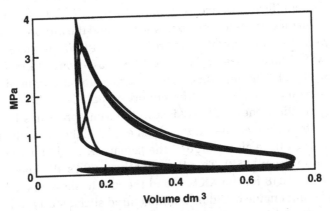

Figure 1.9. Indicator cards generated by ESP for a CFR engine ($b = .0825$ m, $S = 0.114$ m, with a compression ratio of $r = 6.0$). The spark advance is set respectively at $0°, 18°, 26°, 39°$ BTC (before top center). Note that the peak pressure decreases as the spark is retarded.

twin-cam racing engines we have the following figures, where the S/b ratios are averages for the indicated year (from [17]):

Year	S/b
1912	1.91
1922	1.51
1932	1.27
1941	0.89

which brings us to an essentially modern value. It is clear that a low value of the stroke:bore ratio will result in a higher value of output:displacement, so it is difficult to see why the early designers were so enamored of large values of the ratio. The British taxed road vehicles by piston area, so there was considerable incentive to keep the piston area of passenger cars small; this may have carried over to racing engines by force of habit, although it seems unlikely. Higher power:displacement and shorter strokes mean higher engine speeds, and early bearings might be thought to have problems with this. However, most of these racing engines had ball or roller bearings. The only explanation I can think of is piston crown temperature. Smaller bores (relative to stroke) mean lower piston crown temperatures.

Regarding the typical values of specific output, it must be remembered that the 1997 cars included in Figure 1.10 are performance cars, the only kind of interest to the readers of *Road and Track*. The values of power per displacement given are certainly high (except for the Mustang which, despite its enviable reputation in the marketplace, evidently has a plain vanilla engine), but they are not representative of the domestic/import fleet. In 1974, when the Organization of Petroleum Exporting Countries began to control crude oil prices, the U.S. government began to encourage the U.S. automobile industry to modify its designs to reduce the nation's dependence on foreign crude oil. Substantial improvements in Corporate Average Fuel Economy were achieved by the 1987 model year [4], and the CAFE has not changed significantly since then. In [40] we can find the corporate average power per unit displacement for the domestic fleet and the import fleet. From 1978 to 1986 the average of the domestic fleet rose from 21 kW/L to 30 kW/L (28 HP/L to 40 HP/L), while the average of the import fleet rose from 34 kW/L to 42 kW/L (46 HP/L to 56 HP/L). The imports had been under pressure from high fuel costs at the pump for some years due to different taxing policies, which explains their better starting position; they were subject to the same pressures from OPEC, as well as a desire to remain competitive, which explains their parallel increase.

On Figure 1.11 I have also included the Jaguar XK double-overhead cam engine, in 3.8 L form (from 1958). This is one of the world's all-time great engines, first introduced in 1948 for the XK120, and the first to show that a DOHC (Double Over Head Cam) engine could be manufactured successfully in large volume. This engine was used in various models, ending with the XJ6, during a nearly forty year period [89]. When the engine was first brought out, it had a nominal displacement of 3.4 L and a power output of 45.5 kW/L (61 HP/L) at a piston speed of 19.5 m/s (3890 ft/min), with a specific power of 0.484 kW/cm^2 (4.17 HP/in^2).

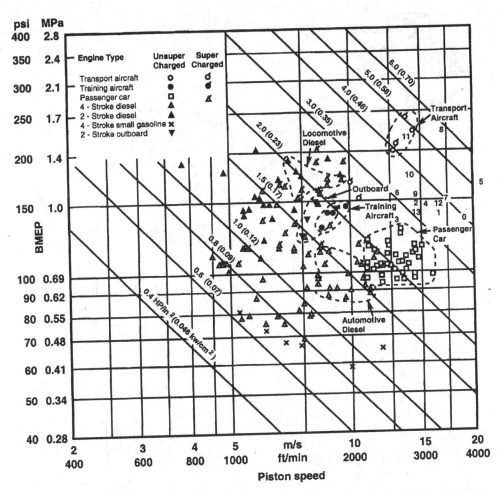

Figure 1.10. Rated brake mean effective pressure, piston speed and power per unit piston area. Published rating, 1954 (*Automotive Industries, Diesel Power*). Aircraft engine ratings are maximum ratings (except takeoff) at either sea level or rated altitude. Diesel and small gasoline engine ratings are continuous maximum others are maximum [94]. I have added values for performance cars for the 1997 model year, taken from *Road and Track*. The figures following the model designation are the specific power in kW/L [HP/L]. 1 = Cadillac Catera (51 [68]); 2 = Chrysler Sebring (50 [67]); 3 = Ford Mustang (35 [47]); 4 = Ford Taurus (52 [69]); 5 = Honda Prelude (67 [90]); 6 = Infiniti Q45t (49 [65]); 7 = Jaguar XK8 (54 [73]); 8 = Mitsubishi Eclipse (78 [105]); 9 = Pontiac Grand Prix (47 [63]); 10 = Porsche Boxter (60 [81]); 11 = Saab 900 Turbo SE (69 [93]); 12 = Saturn SC2 (49 [65]); 13 = Subaru Legacy (47 [63]); O = Jaguar XK 3.8 L 1960 (43 [58]). After [94]. Copyright © 1968 and new material © 1985 by The Massachusetts Institute of Technology.

This is exceptional, putting it at the right-hand edge of Figure 1.10. The stroke:bore ratio was 1.28, a little high.

We will discuss the Mach index and volumetric efficiency in Chapter 2; very briefly, the flow through the inlet valve aperture can reach sonic speed at high piston speed, called choking. This prevents the piston from pulling any more charge into the cylinder and the volumetric efficiency drops. When this happens depends on the valve open area. Here we note in passing that the lift of the intake

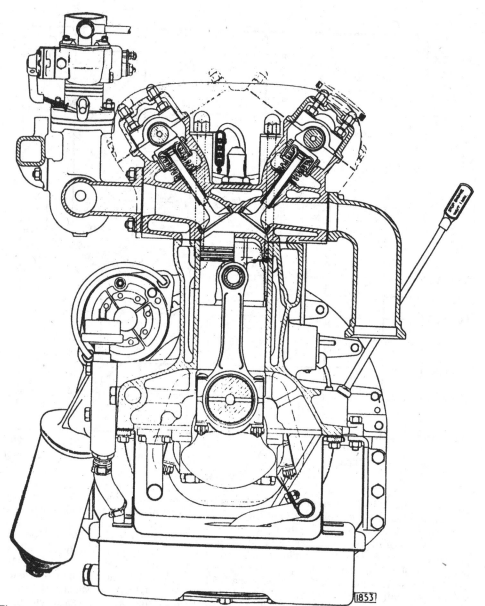

Figure 1.11. The Jaguar XK engine, first used in the XK120 around 1948. At first it was 3.4 L, then the stroke was reduced to produce a 2.4 L version, then the bore of the original version was increased (with the original stroke) to give 3.8 L, and finally, the block was redesigned to give an even larger bore to produce 4.2 L [89]. This is the 2.4 L version. Reproduced by kind permission of Jaguar Cars, England.

valves, divided by their diameter, is about 0.21, which we will find is a little less than optimum (which we will show is 0.28). This gives a Mach index of 0.64 at this piston speed, for a volumetric efficiency of about 0.8.

In order to produce a smaller engine, in 1955 the stroke was decreased substantially (keeping the same bore) to give it a displacement of 2.4 L. This is fairly cheaply done by making a new crankshaft. The stroke:bore ratio is now 0.92, more respectable. To go with the decreased stroke, the piston speed was decreased at the same time by decreasing the valve lift (a new camshaft, but otherwise the cylinderhead was unchanged). We will return later to examine this; by decreasing the valve lift, the designer had increased the Mach index – that is, he had decreased the piston speed at which the flow in the valve aperture became choked. The Mach index is roughly 0.6 at this piston speed, for a volumetric efficiency of about 0.83. Now the engine produced 36.5 kW/L (48.3 HP/L) or 0.276 kW/cm^2 (2.38 HP/in^2) at a piston speed of 14.7 m/s (2930 ft/min).

This is a very good smaller engine that remained in use until 1969 in the cheapest version of the 2.4 L Mk 2 and the similar 240. Limiting the valve lift, and consequently the piston speed, assured the engine a long, reliable life. If the lift had not been reduced the maximum power would have remained about the same with the decreased displacement, which would have made the purchasers of the more expensive 3.4 L version quite unhappy.

In a further modification, keeping the same stroke and valve lift as the 3.4 L, the bore was increased. The power per displacement dropped somewhat to 43 kW/L (58.2 HP/L), or 0.462 kW/cm^2 (3.98 HP/in^2), at the same piston speed of 19.5 m/s (3890 ft/min), probably because the Mach index is a little higher (now 0.70, for a volumetric efficiency of 0.79) – the engine is having trouble breathing. Only ten horsepower were gained, which presumably had a marketing advantage; from an engine designer's point of view, the 3.4 L version was a better engine with higher specific output. In any event, the position of the engine on Figure 1.10 is quite exceptional for the period and quite competitive with the 1997 engines. Note that the XK engine has only two valves per cylinder, but as we shall see, their position in a hemispherical combustion chamber allows them to be as large as possible.

In fact, the engine block was redesigned in 1964 to allow a larger bore of 92.07 mm, taking the displacement up to a nominal 4.2 L, and the engine was used in the E-type, the MkX and the 420 in this form until the early '70s. Using this block, a short stroke version (2.8 L) and a smaller bore version (3.4 L) were produced; the latter was used in the XJ6 until 1986.

For comparison, let us look at a new engine which has just been designed by KIA Motors Co. Ltd., in Korea (described in [64]). In general configuration, it is remarkably similar in cross-section to the Jaguar XK, but with smoother gas flow in the manifolds. Note the modification of the oil pan to allow installation of the engine at an angle. This engine has four valves per cylinder, which are opened essentially to the optimum lift (0.241) for a Mach index of 0.479. Peak power is achieved at a piston speed of 17.4 m/s (3480 ft/min), and this peak is 48 kW/L (64.4 HP/L), or 0.419 kW/cm^2 (3.61 HP/in^2). The *bmep* is 0.958 MPa (139 psi). The stroke:bore ratio is 1.07, which is essentially square. (Square is boy racer jargon for a stroke:bore ratio of unity.) These numbers place it just to the right of 1 and below 7 on Figure 1.10, roughly where the XK engine could have been with a better stroke:bore ratio. Like other modern engines, this one has multi-point fuel injection, rather than a carburetor.

Reading [64] is salutary; the authors are far more concerned with material choice for specified performance, fabrication techniques for minimum cost, gasket design and the engine management system choice, than they are with the

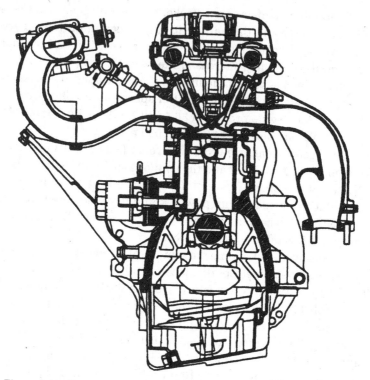

Figure 1.12. Cross-sectional view of new KIA T8D DOHC (Double Over Head Cam) engine [64]. Reprinted with permission from 970917 © 1997 Society of Automotive Engineers, Inc.

things that concern us here. This underscores what I have said before; this is old technology, and we know how to design a good engine from a fluid dynamic/thermodynamic point of view (the principles in this book). Now the problem is to make it of materials that will achieve the potential in the fluid dynamic/thermodynamic design, and make it durable and cheap to manufacture. The only area of physics to which the authors give significant space is the design of the induction system for inertial ram tuning; that is, the design of the intake manifold to take advantage of the inertia and compressibility of the gases to maximize the amount of charge that enters the cylinder. This is still an area in which a little creativity can be exercised, although the principles are well understood. We will deal with intake/exhaust tuning in chapter two.

1.9

POWER EQUATION

We can approach an analysis of the engine also by examining the energy input contained in the fuel. Let Q be the lower heating value (specific calorific value) per unit mass of the fuel. The heat of reaction differs, depending on whether the water in the products is a liquid or a gas. The lower heating value corresponds to water vapor in the products. If $\dot{m}_f$ is the mass flow rate of the fuel, then we can write immediately

$$P_i = \eta_i \eta_c \dot{m}_f Q \qquad (1.11)$$

where P_i is the indicated power, i.e., in the cylinder, not including the effect of friction and some auxiliaries, η_c is the combustion efficiency, and η_i is the indicated efficiency. The combustion efficiency is often combined with the indicated efficiency [94], but here we are keeping it separate. We can move the consideration back to the flywheel and use the brake power P_b by including the mechanical efficiency η_m:

$$P_b = \eta_m \eta_i \eta_c \dot{m}_f Q \qquad (1.12)$$

η_m includes the effects of friction (including turning the camshaft), pumping losses and some auxiliaries (at least the oil pump).

A fuel air mixture will burn reliably in a limited range around stoichiometric. Stoichiometric comes from a Greek word Stoichios, meaning element, and it refers to the chemically correct mixture, in which there is just enough oxygen to burn all the fuel. This ratio, for regular gasoline and air, is $F = 6.76 \times 10^{-2}$ fuel:air, or $1/F = 14.8$ air:fuel, and for premium gasoline and air is $F = 6.80 \times 10^{-2}$ fuel:air, or $1/F = 14.7$ air:fuel [99].

In the lean-burn engine, the homogeneous charge in the cylinder will burn reliably up to air:fuel ratios in the neighborhood of 20, if the turbulence is strong enough. In the stratified charge engine the fuel is confined by various clever techniques to a small fraction of the combustion chamber so that the mixture *in that region* is essentially stoichiometric, and hence will burn reliably, even as the mixture averaged over the entire combustion chamber is substantially leaner than stoichiometric (perhaps 50 air:fuel), because large parts of the combustion chamber contain very little fuel. We will return to this when we discuss pollution.

We can introduce the fuel:air ratio F and the mass flow rate of the air $\dot{m}_a$ in Equation 1.12 and write

$$P_b = \dot{m}_a F Q \eta_i \eta_c \eta_m \qquad (1.13)$$

Now, let us think about the mass flow rate of the air. If air were an incompressible, inviscid medium, then the displacement volume of the engine would be filled with air each two revolutions (for a four-stroke cycle engine) or every revolution (for a two stroke cycle engine). Air, however, is not an incompressible, inviscid medium, and the mass of air which enters the engine is somewhat less, depending on engine speed. One factor is the viscous (turbulent) pressure drop involved in getting the air through the manifold to the cylinder. This results in the air density in the cylinder being below the air density at the entry to the manifold, and hence the mass is somewhat less. The other major factor is the occurrence of shocks in the valve opening at high enough piston speed, choking the flow and also resulting in a substantially decreased density in the cylinder, and hence a reduced mass flow. We can take both these factors into account by introducing a volumetric efficiency η_v, the ratio of the actual mass flow to the mass flow that could be achieved with an inviscid, incompressible fluid:

$$\dot{m}_a = \rho_i S A_p n \frac{N}{X} \eta_v = \rho_i V_d \frac{N}{X} \eta_v \tag{1.14}$$

where ρ_i is the air density at the entry to the manifold (but after a turbocharger or supercharger if present), A_p is the area of a piston, n is the number of cylinders, N is the rotational speed (rps), X is the number of revolutions per power stroke for each cylinder, and V_d is the displacement.

Finally, this can be combined with Equation (1.13) to give

$$P_b = \eta_i \eta_c \eta_m \eta_v \rho_i V_d \frac{N}{X} F Q \tag{1.15}$$

In Section 1.10 we will show how this equation can be used to suggest design possibilities.

1.10
INFLUENCE ON DESIGN

Equation (1.15) is a master equation which will be very useful in design. Each term in the equation suggests some way in which the performance of the engine can be improved. A few possibilities follow. In many of these possibilities terms are mentioned that may not be familiar to you: for example, "separation, boundary layers, and Mach number" in the first entry. These terms will be explained in the text when each subject receives a full discussion.

- η_v: This is influenced by separation, boundary layers, and Mach number. It can be improved by changes in valve timing, cam modifications, porting, intake and exhaust tuning, and changes in the number of valves per cylinder.
- ρ_i: The inlet density can be increased by supercharging, turbocharging, intercooling, use of a fuel with a higher heat of vaporization, and water injection. It can also be increased if the engine is unthrottled, as in Gasoline Direct Injection (GDI).
- V_d: As we have already mentioned, there has been a historical trend to smaller V_d and larger N/X associated with decreases in S/b.
- N/X: See above. In addition, there is the possibility of changing to a two-stroke engine (multiplying the power output by a factor of roughly 1.5), if it can meet the Federal pollution standards. As we have mentioned, Daiwoo in Korea claims to have a blower scavenged two-stroke that meets the pollution standards.
- η_i: The compression ratio can be raised and ceramic coatings for the piston crown can be considered to reduce heat loss. The geometry of the combustion chamber can be changed, along with the number of cylinders for a given displacement.
- η_m: Reduction in piston speed is the main factor here, which requires a smaller S/b ratio. Some racing engines use a single piston ring to increase

η_m at a given piston speed. Low-tension piston rings are also a possibility. There are proprietary low-friction coatings, but I do not believe they work for long; the shear and temperature rapidly degrade the molecules.

Several of the terms used in this itemization are probably unfamiliar. In what follows, we will discuss in detail each of these possibilities, and define the terms.

1.11
bmep AGAIN

If we take Equation (1.5) for the brake power

$$P_b = bmep \, V_d(N/X) \tag{1.16}$$

and compare it with Equation (1.15), we can immediately write

$$\eta_i \eta_m \eta_v \eta_c \rho_i F Q = bmep \tag{1.17}$$

Now, we can see why the value of *bmep* reflects only genuine differences in design – it is affected only by the efficiencies and the inlet density. Engines of completely different characteristics (such as size, speed, and so forth), but equal efficiencies, will have equal values of *bmep*.

F is the fuel:air ratio; at stoichiometric, the air:fuel ratio is 14.89, so that $F = 1/14.8 = 6.76 \times 10^{-2}$. At full power engines operate rich, up to perhaps 1.2 times this value. At such a value η_i falls somewhat, because the properties of the gases in the cylinder are not the same. η_c also falls because not all of the fuel in the mixture is burned. If ϕ (the equivalence ratio) is the ratio of F to the stoichiometric value, then we may roughly parameterize this effect from data given in [94] for a CFR (Coordinating Fuel Research) engine. This parameterization will be different for other engines, but the order-of-magnitude of the effect is similar (the CFR engine is quite atypical, but has provided an enormous amount of data). Roughly

$$\frac{\eta_i \eta_c}{\eta_{ac}} = 0.921 - 0.327\phi \tag{1.18}$$

[94] so that $\eta_i \eta_c / \eta_{ac} = 0.594$ at stoichiometric, while it is only 0.528 at $\phi = 1.2$, a decrease of 11%. However, because the fuel:air ratio is up 20%, the product is up about 7%.

The product $\rho_i F Q$ has the dimensions of a pressure. We may evaluate it for ρ_i atmospheric, and F stoichiometric; then it will represent a value of *bmep*, far above anything practically attainable, because it would correspond to efficiencies of unity. We cannot call it a theoretical maximum, however, because ρ_i could be increased over atmospheric, and F could be increased over stoichiometric. If we take $\rho_i = 1.2$ kg/m^3 and $Q = 42.7$ MJ/kg ([99]) for regular (43.5 MJ/kg for premium), we obtain 3.44 MPa = 499 psi, a number that is worth remembering. Now, consider: a CFR engine with $r = 8$ will have $\eta_{ac} = 0.565$, and at stoichiometric fuel:air ratio (using Equation 1.18) would have $\eta_i = 0.336$. If we take as typical values (we will discuss these in detail later) $\eta_m = 0.85$, and $\eta_v = 0.83$, we get a typical value for $bmep = 0.336 \times 0.83 \times 0.85 \times 3.44 = 0.815$ MPa $= 118$ psi. If we run rich for maximum power, we can produce another 7%, as we have seen, to bring the $bmep$ up to 0.872 MPa $= 126$ psi. However, the inlet air is warmer than the ambient by perhaps $15°C = 27°F$, and the pressure is reduced, by heating in the manifold and the pressure drop required to produce the flow in the manifold (we will discuss this later). The result of this is that the density is only 0.86 of the ambient value [94] (on a standard day, with an ambient temperature of $20°C$), giving a final value of 0.75 MPa $= 109$ psi.

This helps to explain why 1954 passenger cars in Figure (1.10) have values of $bmep$ clustered around 0.76 MPa $= 110$ psi. It is quite difficult to produce much higher values. The values of η_i I have been quoting correspond to CFR research engines that have combustion chambers like cylindrical pill-boxes, with the valves in the head. If we consider more efficient designs of combustion chamber (such as hemispherical), which have less surface per unit volume, and hence lose less heat, we can push the value of η_i up by perhaps 10% at most to give 0.83 MPa $= 120$ psi. By tuning the intake manifold, we can push up η_v by 20%, to give 0.99 MPa $= 144$ psi. Finally, if the intake is tuned the inlet density will be higher by about the same percentage as the volumetric efficiency, for another 20%, to give 1.2 MPa $= 173$ psi. This is something like a practical upper limit for normally aspirated spark-ignition engines operating on normal fuels, as can be seen by an examination of Figure (1.10), although some small improvement is also possible in η_m. It is clear that to go from 0.75 MPa $= 109$ psi to 1.2 MPa $= 173$ psi requires great cleverness and attention to detail. It is a tribute to the manufacturers that between 1951 and 1984 the average $bmep$ rose from about 0.69 MPa $= 100$ psi to about 1.0 MPa $= 150$ psi.

Racing engines can achieve considerably higher values of $bmep$ because they run on exotic fuels. For example, a common fuel (when there are no restrictions on fuel) is 5% nitrobenzine, 80% methanol and 15% acetone (for solubility) [95]. Fuel dragsters (see Figure 2.40 in Section 2.14) run on a fuel like this, in addition to being supercharged. The great advantage of such a fuel comes from two factors: methanol has a very large enthalpy change between liquid and vapor, of order 1.1 MJ/kg, which results in chilling the incoming charge; and, methanol is very resistant to knock. (In addition, nitrobenzine and methanol carry oxygen, and consequently increase the energy produced per mass of air.) As a result, compression ratios in the neighborhood of 16:1 can be used in an engine running on such a fuel. This results in much higher values of η_i. At $r \doteq 16$, $\eta_{ac} = 0.67$. If we take $\eta_i/\eta_{ac} = 0.528$, which would be approximately true for a CFR engine at $\phi = 1.2$, we can guess a value of $\eta_i = 0.354$. This fuel has a $Q = 20.9$ MJ/kg, a relatively low value, and a correspondingly large $F = 0.150$. For the product $\rho_i Q F$ we obtain a value of 4.51 MPa $= 655$ psi (substantially higher than the value of 3.44 MPa obtainable with conventional fuels, due primarily to the nitrobenzine that brings its own oxygen). Taking $\eta_v = 0.83$ and $\eta_m = 0.85$, we obtain as an estimate for $bmep = 1.1$ MPa $= 159$ psi. With tuning, it is possible to achieve in the neighborhood of 1.4 MPa = 200 psi, and normally aspirated racing engines running on these fuels do routinely achieve figures close to that [95]. Our estimate of η_i/η_{ac} is probably somewhat high; as r increases, the geometry of the combustion chamber changes, the volume necessarily decreasing while the surface area stays roughly constant. At $r = 16$, the combustion chamber is very thin:the volume:area ratio (a characteristic length) is of the order of the clearance height (the distance between the top of the piston at TC and the underside of the combustion chamber top). This increases the heat loss, and consequently decreases η_i.

In the chapters that follow, we will discuss these various possibilities.

1.12
SOME MORE THERMODYNAMICS

We must talk in a general way about what goes on in the cylinder to set the stage for the discussions that will follow. In Chapter 2 we will talk about the physics of getting the mixture into and out of the cylinder. In Chapter 3 we will talk about the physics of heat transfer in the cylinder. In Chapter 5 we will talk in detail about the flow and turbulence in the cylinder. Finally, in Chapter 8 we will examine in detail how ESP calculates the flow into the cylinder and the temperatures, pressures and turbulence in the cylinder, before and after combustion.

1.12.1 Turbulence and Flow in the Cylinder

The mixture in the cylinder during the compression stroke generally is far from uniform. Depending on the design of the engine there may or may not be large-scale organized rotation of the flow, either swirl (with the axis aligned with the axis of the cylinder) or tumble (with the axis at right angles to the axis of the cylinder). See Figures (5.3) and (5.4). This organized motion is often called coherent – to distinguish it from incoherent, chaotic turbulent motion. During the intake stroke, air rushed in through the intake valve, and this jet of air was turbulent – that is, unsteady and inhomogeneous. After the valve closed, most of the mean-flow energy in this jet was also converted into turbulence. Both the turbulence and the large-scale rotation attempt to mix the contents of the cylinder. By the time of ignition – that is, near TC – if the swirl or tumble is not too strong this turbulence/mixing is relatively homogeneous on the larger scales [47], although analysis of the composition of the products immediately after ignition indicates that there may be still considerable inhomogeneity on the smaller scales. If an effort has been made to stratify the charge – that is, to prevent the charge from being homogeneous, keeping the fuel from mixing with all the air – then the charge will not be homogeneous even on the large scales. Stratification is usually associated with strong swirl or tumble. Squish (if part of the piston crown at TC is very close to part of the combustion chamber roof) also introduces coherent motion as well as more turbulence.

1.12.2 Heat Transfer

The calculation of what is happening in the cylinder depends on a calculation of heat transfer. This requires a knowledge of the temperature of the various surfaces in the cylinder, the areas of the surfaces exposed to the burned and to the unburned gases (which, of course, have different temperatures), and the gas temperatures. Most important, we need to know the gas phase heat transfer coefficient, which is primarily dependent on the turbulence in the cylinder. We will discuss this at much greater length in Chapter 3.

1.12.3 Chemical Reaction

Although a laminar flame (reaction zone) is very thin, typically of order 0.5 mm [21], the flame sheet in an engine combustion chamber is heavily distorted by the turbulence, folded and crumpled and wrinkled so that the zone in which reaction takes place is quite thick [47]. This is referred to as a flame brush. As the flame brush proceeds across the combustion chamber, a chemical change takes place, converting the mixture of fuel, fuel vapor, oxygen, nitrogen, water vapor and trace amounts of other species into (for example) C, CO, CO_2, H, OH, H_2, H_2O, N, NO, NO_2, N_2, O and O_2. The concentration of all these products depends on the temperature; throughout most of the expansion/power stroke, the temperature is high enough for chemical equilibrium to exist. During the power stroke, the turbulence is strongly suppressed, and the mixing largely stops. As the temperature of the mixture falls the reaction rates fall very rapidly. The reaction rates contain what is called an Arrhenius factor [8] of the form $e^{-E/RT}$, where E is a constant called the activation energy, and R is the Universal Gas Constant. As T drops the value of the Arrhenius factor drops very rapidly. At about the time the exhaust valve opens, the temperature has fallen low enough so that the composition becomes frozen – that is, the reaction rates become very low at these low temperatures and significant changes happen only on a much longer time scale (days, or even weeks). We will return to this when we discuss pollution. It is this freezing of the composition that results in the release of various oxides of nitrogen into the atmosphere.

The way in which the flame brush proceeds across the combustion chamber depends on the turbulence and the coherent motions. If the coherent motions are not very strong the flame brush will proceed as an approximately spherical front from the spark plug. It takes a relatively strong coherent motion to distort the flame brush significantly [47]. The area of the flame brush and its speed determines the rate at which reactants are converted into products and heat is released. The area of the flame brush is determined by the geometry of the combustion chamber and the placement of the spark plug; this is a calculation that can be carried out for each combustion chamber and spark plug location, and the results of such a calculation are required by ESP (see, e.g., [78]). The speed of the flame brush is determined largely by the turbulence velocity, which carries portions of the flame sheet forward and back.

1.12.4 STANJAN, ESPJAN and ESP

If we are at a temperature high enough for chemical equilibrium to be a reasonable assumption, and if we can assume thorough mixing, the equilibrium concentration of the various components can be computed. The properties of the reactants and the products can be computed if we know the composition. These properties can be determined from the JANAF [74] tables. These tables allow calculation of chemical equilibrium states of complex mixtures using computer programs developed for this purpose, such as STANJAN. If the fuel, oxidizer and allowable species products are specified, as well as the temperature and pressure, the properties of the reactants and products can be determined. To the extent that the reactants and products can be regarded as perfect gases, the internal energy and enthalpy are not functions of pressure, and are only functions of temperature.

Also available through the ESP website is the program ESPJAN. When the fuel, oxidizer and products are specified, and the fuel:air ratio is specified and a pressure is given, ESPJAN calculates a table giving the values of the enthalpy h and the product Pv, where P is the pressure and v is the specific volume (the inverse of the density), for the reactant mixture and the product mixture, as functions of temperature at intervals of 100 K from 200 K to 4,900 K. This table is then used by ESP to calculate what is happening in the cylinder. This assumes that the reaction products are in chemical equilibrium, and that the effect of pressure on the equilibrium composition is negligible, which are reasonable approximations for this level of analysis.

1.12.5 Heating Values and Enthalpy

We are interested in the enthalpy because according to the first law of thermodynamics the difference in enthalpy between the reactants and the products equals the energy released in a reaction at constant pressure [47]. The chemical reaction in the flame brush occurs at approximately constant pressure; the pressures on the two sides of the flame brush are equal. The heating value of the fuel we have designated by Q, which is found in handbooks, is actually the difference between the enthalpy of the reactants and the enthalpy of the products at a standard reference temperature, usually about 300 K, assuming that the water in the products is in vapor form. It is a matter of convention to assign this to the fuel.

1.13
PROBLEMS

1. • Show that the average piston speed for a four-stroke engine $\overline{V}_p = 2NV_d/nA_p$, where $N=$ engine speed, $V_d=$ total displacement, $A_p=$ piston area of one cylinder, and n is number of cylinders.
   • Then show in the equation for the power of an engine that N and V_d can be eliminated in terms of A_p and $\overline{V}_p$.

2. Working with ESP, using no manifold, and any convenient configuration, change the point of ignition to 30, 20, 10 and 0 CAD BTC, and determine the peak pressure and the CAD at which peak pressure occurs. Discuss implications for knock.

3. Working with ESP, using no manifold, and any convenient configuration, modify the heat transfer model by setting all the Stanton numbers (1–7) to zero. Now determine the indicated efficiency. Compare with indicated efficiency with Stanton numbers given in lumcfrx.ess at same ignition point. Discuss.

4. • Derive the equation that describes the piston motion in the cylinder from a slider-crank model. (Let piston travel $=x$, crank angle $=\theta$, half stroke $=R$, connecting rod length $=L$.) Assume $x=0$ at $\theta=0$ at TC. Use geometry.

   • Given an engine with bore $b=85$ mm, stroke $S=100$ mm, and $L=185$ mm, use the equation found above to plot the percent total displacement versus crank angle for 180 degrees of crank rotation. At TC, the percentage total displacement $=0$. You only need to plot 10 points to cover the full 180 degrees.

5. At 2000 rpm the engine of your car has an indicated efficiency of $\eta_i=0.304$ (corresponding to a compression ratio $r=8.5$) at an equivalency ratio (ratio of fuel:air ratio to stoichiometric) of $\phi=1.2$ (this rich mixture gives maximum power). Note that the stoichiometric air:fuel ratio is 14.9. Your engine has a volumetric efficiency $\eta_v=0.83$ and a mechanical efficiency $\eta_m=0.85$. The heating value of the fuel is $Q=42.7$ MJ/kg. The air density is $\rho_i=1.2$ kg/m^3. What is the *bmep*? (Hint: use the two expressions for brake power).

6. You are designing an engine. You want to achieve 37 kW/L at a *bmep* of 1 MPa and a piston speed of 15 m/s. What stroke should you use?

BREATHING EXERCISES

INTRODUCTION

When the flow velocity through an orifice reaches the local speed of sound, a change in the pressure downstream of the orifice can no longer be communicated to the flow upstream of the orifice. This is because the message would have to travel upstream as a pressure wave at the speed of sound, and it cannot do this if the stream is flowing against it at this speed. The flow is called *choked*, and the mass flow rate cannot be increased beyond this point. We call the ratio of the local fluid speed U to the speed of sound a, the Mach number $M = U/a$. When the $M = 1$ at the orifice, the flow is choked (see, for example, [31]).

It is at first surprising that sonic conditions, or supersonic flow, have anything to do with an automobile engine. It turns out, however, that the flow through the inlet valve becomes choked under normal operating conditions, and this is one of the most serious limitations on the performance of an engine. The design of the valve gear is dictated largely by the need to avoid choked flow throughout the desired performance range. The desired upper end of the performance range determines when the flow through the inlet valve will become choked.

In order to control the engine speed at which the flow becomes choked, the engine designer must control the size and lift and number of the inlet valves, and this influences the combustion chamber shape, as well as the nature of the valve gear (for example, pushrod, overhead cam, double overhead cam).

2.2
FLOW THROUGH THE INLET VALVE

Let's talk first about the nature of the flow through an inlet valve. In Figure 2.1 (from [1]) we show patterns of flow through a sharp-cornered inlet valve, typical of valves in common use. Notice the flow separation; at low lift there is a small separation bubble just after the sharp inner corners, but the flow reattaches and fills the gap; at intermediate lift, the flow reattaches on the seat, but not on the valve; at high lift, the flow separates from both corners, and forms a

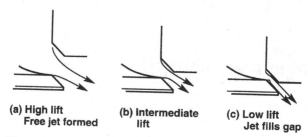

(a) High lift
Free jet formed

(b) Intermediate
lift

(c) Low lift
Jet fills gap

Figure 2.1. Patterns of flow through a sharp-cornered inlet valve [1]. Reproduced by kind permission of W. J. D. Annand.

free jet. We will need to carry this picture of the valve flow in mind, both for our discussion of choking, as well as our later discussion of turbulence in the cylinder. Separation occurs in many flows of practical interest. It often occurs at a sharp edge, although not always. Here, the fluid cannot negotiate the sharp corners.

Notice, that beyond a certain point, lifting the valve more does not increase the flow, since the flow is not filling the area available to it. For low lift, the smallest area is the area between the valve and the seat. However, at high lift, the smallest area is the area of the valve port. Let us define some terms. Figure 2.2 shows a valve and port with various distances identified. The area πDL is referred to as the curtain area. It is clear that this is not the actual area available to the flow, even for small lift. The gap normal to the valve seat is $g = L \cos\theta$, and the flow area A_f available is something like

$$A_f = \pi \frac{D + D_i}{2} g = \pi \frac{D + D_i}{2} L \cos\theta \tag{2.1}$$

In the case of the Jaguar XK engine, for example, $D_i = 38.1$ mm and $D = 44.45$ mm, so that the ratio $D_i/D = 0.857$, and this is quite typical (observed values ranging from 0.812 to 0.930 [1]). In the Jaguar XK engine, $\theta = 45°$, so that $\cos\theta = 0.707$.

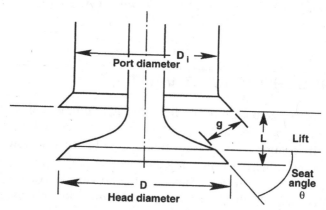

Figure 2.2. A typical inlet valve with various distances identified.

The actual flow area is

$$A_f = \pi DL \frac{1 + D_i/D}{2} \cos \theta \qquad (2.2)$$

The factor $\frac{1+D_i/D}{2} = 0.929$ for the XK engine, and $\cos \theta = 0.707$, so that $A_f = 0.656\pi DL$. Nevertheless, the curtain area is typically used as a measure of flow area in the literature, because it is easy to determine. In the program ESP, the flow area must be specified, and this must be the actual flow area A_f, Equation 2.2.

When the valve lift is large, the smallest area is the port area, $A_m = \pi D_i^2/4$ (m for minimum). However, D_i is often difficult to determine, while D is easy. Hence, consider the relation

$$A_m = \pi \frac{D^2}{4} \frac{D_i^2}{D^2} \qquad (2.3)$$

For the XK engine, the factor $D_i^2/D^2 = 0.734$. The two areas are equal when

$$\pi DL \frac{1 + D_i/D}{2} \cos \theta = A_f = A_m = \frac{\pi D^2}{4} \frac{D_i^2}{D^2} \qquad (2.4)$$

or, cancelling common factors and solving for L

$$L = \frac{D}{4} \frac{\left(\frac{D_i}{D}\right)^2}{\cos \theta \frac{1 + \frac{D_i}{D}}{2}} \qquad (2.5)$$

In the case of the XK engine, the factor gives $L = 1.12D/4$. Thus, there is no point in lifting the valve more than this, since it does not increase the area available for the flow. $D/4$ is used as a rough rule of thumb for maximum valve lift.

2.3
THE DISCHARGE COEFFICIENT

There are a number of ways in which the flow through a valve (or any other orifice) can be evaluated. One of the most direct is the discharge coefficient, which is used in the ESP program.

We can use isentropic compressible flow theory to calculate the mass flow rate through an imaginary frictionless orifice of the same area as the valve opening, operating with the same pressure difference, between infinite reservoirs. Let us call this $\dot{m}_f^i$ for ideal. We have to use compressible flow theory, because this flow will approach or exceed the speed of sound. Now, the valve opening is hardly an ideal frictionless orifice, since it has sharp edges that induce separation. Hence, the mass flow rate will be less than the value from isentropic compressible flow theory. Let us call the real mass flow rate $\dot{m}_f$. It is usual (see, for example [31]) to define a discharge coefficient C_D as the ratio of the real to the theoretical mass flow rate:

$$C_D = \frac{\dot{m}_f}{\dot{m}_f^i} \qquad (2.6)$$

This is the discharge coefficient that is used in ESP.

There are other ways of defining a discharge coefficient. For example, [1] introduce the concept of the effective area A_f^e. This is the area of an imaginary frictionless orifice which would produce the real mass flow rate $\dot{m}_f$. Since, however, the mass flow rate through an ideal orifice is proportional to the area (the coefficient of proportionality is dependent on the upstream and downstream conditions in the gas), the discharge coeffcient is also proportional to the area ratio:

$$C_D = \frac{A_f^e}{A_f} \qquad (2.7)$$

Finally, instead of using the true area, A_f, we could use in the definition of the discharge coefficient some other area that is easier to measure, such as the curtain area [1]. This will not be the same discharge coefficient, because of the difference in the reference area. Let us call it C_D^a:

$$C_D^a = \frac{A_f^e}{\pi\,DL} = \frac{A_f^e}{A_f}\frac{A_f}{\pi\,DL} = C_D \frac{A_f}{\pi\,DL} \qquad (2.8)$$

We reproduce here from [1] a plot of C_D^a vs. L/D. The three segments of the curve shown correspond to the three regimes in Figure 2.1. The dashed portions of the curve, joining the three segments, represent unsteady states during which the flow is changing from one regime to the other. Note that, up to $L/D = 0.28$ approximately, the curtain area is proportional to the actual area, and only at about 0.28 does the actual area stop increasing when L/D continues to increase. Hence, the drop shown in Figure 2.3 for values of $L/D > 0.2$ is real, and is the result of separation (that the flow does not fill the available area, and hence the effective area falls off). To obtain the values of C_D defined relative to the true flow area A_f, the numerical values on the abscissa of Figure 2.3 must be multiplied by $\pi\,DL/A_f \approx 1/0.656$. It is evident that, until somewhere near $L/D = 0.2$, the values of C_D (relative to the true area) are nearly unity. Hence, it is not until substantial separation has occurred that the (true) discharge coefficient drops substantially below unity.

It is possible to improve the values of the discharge coefficient somewhat [1] by modifying the flow passage just upstream of the valve – rounding the sharp corners, for example. The values can be lowered by a bad placement of

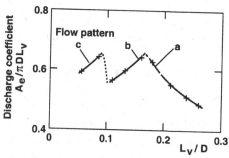

Figure 2.3. Typical variation of discharge coefficient with lift/diameter ratio for an isolated, sharp-cornered inlet valve [1]. Reproduced by kind permission of W. J. D. Annand.

the valve in the cylinder, or a badly designed inlet port. The situation can often be improved somewhat by judicious use of a hand-held die grinder, to smooth and re-shape the port. This is known as porting, and is usually done to high-performance engines. An improvement of C_D^a from 0.6 to nearly 0.7 is possible.

2.4
THE FLOW COEFFICIENT

Still another way to examine the flow through the valve is to compare the actual flow to the constant density, incompressible, inviscid flow that would be present due to the piston motion. This may not be the most convenient comparison fluid mechanically, but it will have the advantage of being easy to determine. Let us define V_p and A_p as the (instantaneous) piston velocity and area ($A_p = \pi b^2/4$), and V_v^i as the ideal (= incompressible, inviscid) gas velocity at the valve opening. Let A_v be the area of the valve, $= \pi D^2/4$. Then if A_f is the actual flow area at the valve opening, we can write

$$V_v^i A_f = V_p A_p \tag{2.9}$$

We can rewrite this as

$$V_v^i = V_p \left(\frac{A_p}{A_v}\right)\left(\frac{A_v}{A_f}\right) = V_p \left(\frac{b}{D}\right)^2 \left(\frac{A_v}{A_f}\right) \tag{2.10}$$

We have introduced A_v because it is a convenient measure of the maximum flow area, although we know that the true maximum area, A_m is actually somewhat smaller (of the order of $0.734 A_v$ as we have seen above). The ratio A_f/A_v is a flow coefficient,

$$\frac{A_f}{A_v} = 0.656 \times 4\frac{L}{D}, \quad L \le 1.12\frac{D}{4} \tag{2.11}$$

using the numerical values for the XK engine, which are typical. For maximum lift,

$$\frac{A_f}{A_v} = \frac{A_m}{A_v} = 0.734 \tag{2.12}$$

again using the value for the XK engine.

Now, we will consider the real flow velocity through the valve, which we will call V_v. We will write

$$V_v = V_p \left(\frac{b}{D}\right)^2 \frac{1}{C_v} \tag{2.13}$$

C_v is a flow coefficient. We expect that C_v will be smaller than the values given in Equations 2.11 and 2.12. The flow coefficient is measured under steady state conditions, with velocities low enough for the flow to be incompressible. Clearly, if the velocity through the valve were in the compressible regime, the flow coefficient would not have a unique value, since the ratio V_v/V_v^i would depend on Mach number. However, all we need is a convenient measure of the velocity through the valve, in order to compute a sort of Mach number, as we shall see in the next section. For this purpose, Equation 2.13 will do very well. We reproduce here a plot of C_v vs. L/D: Notice that the values of C_v are a bit larger than the values suggested by Equations 2.11 and 2.12 until $L/D \approx 0.28$, and a bit smaller thereafter. This means that $V_v < V_v^i$ for $L/D < 0.28$, but $V_v > V_v^i$ for $L/D > 0.28$. The latter is small and probably not significant, representing only our somewhat inaccurate estimate of the flow area available. In the former case, when the lift is small, the losses due to the boundary layers and the separation cause a reduction in density as the flow passes through the valve, reducing the velocity below the inviscid, incompressible value.

2.5

THE MACH INDEX AND VOLUMETRIC EFFICIENCY

Now we are finally in a position to estimate the Mach number of the flow through the valve. We will use our estimate of the velocity V_v, Equation 2.13. Approximate as this is, it is still a little inconvenient, because it is instantaneous – that is, it varies during the engine cycle. For design purposes, it would be useful to have an average expression that gives us one figure for the entire period that the inlet valve is open. In 2.13 we can replace V_p by an average value $\overline{V}_p = 2NS$, where S is the stroke, and $1/2N$ is the time for one-half a revolution of the crank. However, we need to do something about C_v. In Figure 2.5 is a plot of the variation of C_v with crank angle for a typical engine. Note that the valve $L/D = 0.24$ approximately. Increasing the value of L/D will not increase the value of C_v significantly, as can be seen from Figure 2.4. A high-lift cam, for example, can only raise the shoulders of the C_v curve, increasing its value during the opening and closing phase.

We will replace C_v with its average value over the time that the inlet valve is open, and we will call this average value C_i:

$$C_i = \overline{C}_v = \frac{1}{\Theta} \int_{\theta_1}^{\theta_2} C_v \, d\theta$$

$$(2.14)$$

where θ_1 is the angle at which the inlet valve opens (about 15° BTC on Figure 2.5) and θ_2 is the angle at which the inlet valve closes (about 254° ATC on Figure 2.5). $\Theta = \theta_2 - \theta_1$, the total crank angle that the inlet valve is open. On Figure 2.5 this corresponds approximately to 269°. We may estimate C_i, without going to the trouble of carrying out the integration, as approximately one-half the maximum value of C_v (corresponding to a normal cam profile, approximately a cosine curve,

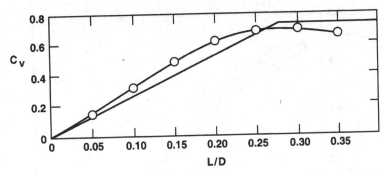

Figure 2.4. Curve of C_v vs. L/D from experimental results [94]. Copyright ©
1968 and new material © 1985 by The Massachusetts Institute of Technology.
We have included the values from Equations 2.11 and 2.12.

as used on ESP). In [94], the value of $C_i \approx 1.45$ L/D is recommended, which is
approximately the same thing. L is now the maximum lift.

Now we have for our estimate for the average gas velocity through the inlet
valve

$$\overline{V}_v = \overline{V}_p \left(\frac{b}{D}\right)^2 \frac{1}{C_i} \tag{2.15}$$

We can make a Mach number using this estimate for the velocity through the valve.
We will call it a Mach index, however, rather than a Mach number, because $\overline{V}_v$ is
such a crude estimate of the actual velocity through the valve; we will denote the
Mach index by Z:

$$Z = \frac{\overline{V}_v}{a} = \frac{\overline{V}_p}{a} \left(\frac{b}{D}\right)^2 \frac{1}{C_i} \tag{2.16}$$

where a is the isentropic speed of sound, 343 m/s at 20°C and 101.4 kPa in dry air
[99], [104]. Between 0% and 100% humidity at 20°C the speed of sound only rises

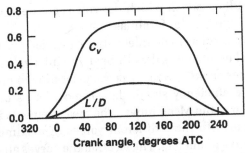

Figure 2.5. Plot of C_v vs. crank angle for a typical
engine [94]. Copyright © 1968 and new material
© 1985 by The Massachusetts Institute of Technology.

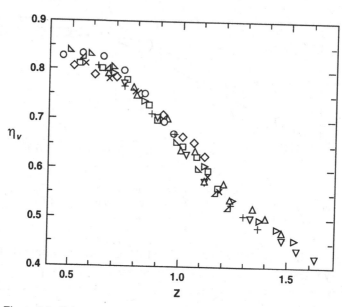

Figure 2.6. Volumetric efficiency vs. inlet valve Mach index [65]. $r = 4.92$; $b = 82.6$ mm, $S = 114$ mm; valve overlap $= 6°$; inlet closes 55° late; $T_i = 322$ K; $T_c = 356$ K; $p_i = 0.931 \times 10^5$ Pa; $p_e = 1.07 \times 10^5$ Pa; $F_R = 1.17$. Inlet pipe length zero. Various maximum lift:diameter ratios.

to 344 m/s [104]. It is essentially only a function of temperature, being proportional to the square root of absolute temperature: $a = \sqrt{\gamma R T}$.

Equation 2.16 assumes that there is only one inlet valve. If there are n inlet valves (presuming they are all open the same amount), there is n times the flow area, so that the velocity through a valve will be $1/n$ of the single-valve velocity. As a result, the Mach index will be given by

$$Z = \frac{\overline{V}_v}{a} = \frac{1}{n} \frac{\overline{V}_p}{a} \left(\frac{b}{D}\right)^2 \frac{1}{C_i} \tag{2.17}$$

It is clear how to handle cases in which the valves are open different amounts, but it is not useful to give general formulas.

Let's return to the case of a single inlet valve. Now, the question is, can we use this Mach index to tell us when the flow through the inlet valve becomes choked? When the flow becomes choked, we expect the mass flow to not increase further, so that the volumetric efficiency should fall. Hence, let us plot the volumetric efficiency as a function of the Mach index. The measurements in Figure 2.6 were made on a CFR engine with three different values of maximum valve lift, and three different inlet valve diameters. The fact that the curves all collapse indicates that the Mach Index is the only relevant parameter. In Figure 2.6, the value of volumetric efficiency at low Z (0.83) is characteristic of this particular engine and valving (the values of L/D ranged from a little below 0.2 to a little above 0.3, but the ratio of valve area to bore area, $(D/b)^2$ ranged only from 0.06 to 0.1 – higher values of η_v would require much larger values of $(D/b)^2$, similar to that of the XK engine, 0.287).

Note that the curve in Figure 2.6 begins to fall at a Mach index of roughly 0.5. Remember that the Mach index is not a Mach number, but simply an estimate of it. The fact that η_v begins to fall when Z is only 0.5 suggests that the instantaneous Mach number is actually reaching a value of 1.0 at some point during the time that the inlet valve is open.

To get an idea of the variation of the Mach number during the time that the inlet valve is open, we could construct a model of the variation of V_p and C_v. This is relatively easy in the middle of the period, but is quite difficult to do realistically near the ends of the cycle, as both quantities are going to zero. Heywood [47] has a realistic estimate, which suggests that the instantaneous Mach number reaches a maximum just after opening, and again in mid cycle. We may expect choking to occur first at these points, and as the Mach index rises, the fraction of the period during which the flow is choked will increase, so that eventually the flow will be choked nearly the entire time the valve is open.

2.6
PARTIAL THROTTLE

Engines are sold (largely) on the basis of their maximum output, but they are seldom driven at maximum output. At partial throttle, the manifold pressure is substantially below atmospheric, and hence, also the inlet density ρ_i at the valve port. We should talk about what effect this has on the volumetric efficiency.

Let us suppose that the pressure in the inlet port, p_i, is below the pressure in the exhaust, p_e. Consider an engine without significant valve overlap (see Section 2.10), that is, one in which the exhaust valve is essentially closed before the inlet valve opens. At the end of the exhaust stroke, the clearance volume (above the piston) is still filled with exhaust gases; these are called the residual gases. These are at the pressure in the exhaust port, p_e. When the inlet valve opens, the pressure in the inlet port, p_i, is below the pressure in the clearance volume, and the residual gases flow into the inlet manifold until the pressure is equalized. When the piston starts down, these gases are drawn back into the cylinder. Since this is not fresh charge, it reduces the volumetric efficiency.

A similar situation occurs if the engine is supercharged or turbocharged, so that the pressure in the inlet port is above the pressure in the exhaust port. Then, when the inlet valve is opened, the fresh charge in the inlet port flows into the cylinder until the pressure is equalized, increasing the volumetric efficiency.

If we assume that the residual gases and the fresh charge have the same specific heats and molecular weights, and are perfect gases, and that the exhaust valve closes and the inlet valve opens at TC, and the equalization of pressure takes place at TC, we can determine a volumetric efficiency for this process [94]:

$$\eta_v = \frac{\gamma - 1}{\gamma} + \frac{r - (p_e/p_i)}{\gamma(r - 1)} \tag{2.18}$$

where $\gamma = c_p/c_v$, the ratio of specific heats.

The estimates from Equation 2.18 compare quite well with measurements in real engines [94]. In making these measurements, it is assumed that Equation 2.18 gives a reduction below the value of the volumetric efficiency that the engine would have if $p_e = p_i$. That is, if we designate η_v^1 as the volumetric efficiency when $p_e/p_i = 1$, we have

$$\frac{\eta_v}{\eta_v^1} = \frac{\gamma - 1}{\gamma} + \frac{r - (p_e/p_i)}{\gamma(r - 1)} \tag{2.19}$$

As an example, if $r = 8$ and $p_e = 100$ kPa (about atmospheric pressure), while $p_i = 37$ kPa (roughly the intake manifold pressure of a healthy engine at idle), then $p_e/p_i = 2.7$, and we find that $\eta_v/\eta_v^1 = 0.83$.

From this analysis, it is clear that the gas velocities through the valve are unaffected by the partial throttle condition. The manifold density is reduced, and the gas passing through the valve may be residual exhaust making a return trip, but the gas velocity analysis of Section 2.5 is still valid. Hence, at partial throttle we expect to begin at low piston speed with a volumetric efficiency which is reduced due to the backflow of residual gases, and which will drop further as piston speed increases, due to the rising Mach index, approximately following Figure 2.6, where now the ordinate should be labeled $\eta_v/\eta_v^{0.5}$, where $\eta_v^{0.5}$ is the volumetric efficiency when the Mach index is equal to 0.5.

As the piston speed increases at partial throttle, the Mach index will increase, the volumetric efficiency will drop and consequently the *bmep* will drop, requiring the driver to open the throttle wider to maintain the same performance.

2.7

THE XK ENGINE

As an example, we can look at the XK engine again. When the engine first appeared as a nominal 3.4 L it had the following specifications:

$$b = 83 \text{ mm}$$
$$S = 106 \text{ mm}$$
$$V_d = 3.442 \text{ L}$$
$$P_b = 157 \text{ kW (210 HP)}$$
$$\text{rpm}|_{maxHP} = 5{,}500$$
$$\text{specific output} = 0.484 \text{ kW/cm}^2 \text{ (4.17 HP/in}^2)$$
$$P_b/V_d = 45.5 \text{ kW/L (61 HP/L)}$$
$$\overline{V}_p = 19.4 \text{ m/s (3,890 ft/min)}$$
$$L \text{ (valve lift)} = 9.53 \text{ mm}$$

$$D_i \text{ (ID valve seat insert)} = 38.1 \text{ mm}$$
$$D \text{ (OD valve head)} = 44.45 \text{ mm}$$
$$L/D = 0.214$$
$$C_i \text{ (estimated)} = 0.31$$
$$Z = 0.639$$

The engine is nearly ideal in this form, although the S/b ratio is a little large. However, the specific output is excellent.

As we said earlier, the engine was downgraded to a nominal 2.4 L for another application by installation of a shorter-throw crank and a lower lift cam. The head was unchanged. The specifications now are:

$$b = 83 \text{ mm}$$
$$S = 76.5 \text{ mm}$$
$$V_d = 2.483 \text{ L}$$
$$P_b = 89.5 \text{ kW (120 HP)}$$
$$\text{rpm}|_{maxHP} = 5,750$$
$$\text{specific output} = 0.276 \text{ kW/cm}^2 \text{ (2.38 HP/in}^2\text{)}$$
$$P_b/V_d = 36.0 \text{ kW/L (48.3 HP/L)}$$
$$\bar{V}_p = 14.7 \text{ m/s (2930 ft/min)}$$
$$L \text{ (valve lift)} = 7.94 \text{ mm}$$
$$D_i \text{ (ID valve seat insert)} = 38.1 \text{ mm}$$
$$D \text{ (OD valve head)} = 44.45 \text{ mm}$$
$$L/D = 0.179$$
$$C_i \text{ (estimated)} = 0.259$$
$$Z = 0.577$$

We see that Z has not decreased much, not enough to cause an appreciable rise in η_v (see Figure 2.6).

In a further modification, the engine was bored out to a nominal 3.8 L, leaving the head, valves, cam and crank alone. The specifications now are:

$$b = 87 \text{ mm}$$
$$S = 106 \text{ mm}$$
$$V_d = 3.78 \text{ L}$$
$$P_b = 164 \text{ kW (220 HP)}$$
$$\text{rpm}|_{maxHP} = 5,500$$
$$\text{specific output} = 0.462 \text{ kW/cm}^2 \text{ (3.98 HP/in}^2\text{)}$$
$$P_b/V_d = 43.3 \text{ kW/L (58.2 HP/L)}$$
$$V_p = 19.5 \text{ m/s (3890 ft/min)}$$
$$L \text{ (valve lift)} = 9.53 \text{ mm}$$
$$D_i \text{ (ID valve seat insert)} = 38.1 \text{ mm}$$
$$D \text{ (OD valve head)} = 44.45 \text{ mm}$$
$$L/D = 0.214$$
$$C_i \text{ (estimated)} = 0.31$$
$$Z = 0.703$$

In fact, Z is a little too high, causing a degradation in specific output, although the total power is increased, which probably had marketing advantages.

The valve timing has also changed between the 2.4 L engine and the 3.4/3.8 L engines. For the 2.4 L engine, the intake opened at 10 CAD (crank angle degrees) BTC and closed 50 CAD ABC, while the exhaust opened 57 CAD BBC and closed 15 CAD ATC. In the 3.4/3.8 L engines, the exhaust timing was left unchanged, but the intake was opened 5 CAD earlier, and closed 7 CAD later. We will see later in Section 2.10 that this will improve the high-rpm performance.

2.8
COMBUSTION CHAMBER SHAPE

We have seen that it is desirable to have the largest possible valve area relative to the bore area at both partial and full throttle, since this will keep the value of Z low. We can increase the valve area by changing the shape of the combustion chamber and by changing the number of valves.

A flat cylinder head, with a combustion chamber the same diameter as the bore, places severe restrictions on the diameter of the valves. We reproduce here a figure giving some representative arrangements of valves in such a cylinder head, from [95].

Notice that the arrangement with three valves, with one inlet and two exhaust, has a higher piston speed for $Z=0.6$ than the arrangement of four valves per cylinder.

It is interesting to ask, in that case why the widespread use of four valves per cylinder at the present time? In the first place, there are other considerations,

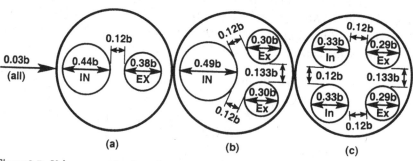

(a)   (b)   (c)

Figure 2.7. Valve area ratios for a flat cylinder head [95].

Arrangement	$\frac{D_i}{b}$	$\frac{D_e}{b}$	$\frac{A_i}{A_p}$	Piston Speed for $Z=0.6$ m/s (ft/min)
a – two valves	0.44	0.38	0.193	14.7 (2900)
b – one in, two ex	0.49	0.30	0.24	18.4 (3620)
c – four valves	0.33	0.29	0.218	16.7 (3300)

For all cases clearance to bore = $0.03b$, space between valves = $0.12b$, except $0.133b$ between two exhaust valves. $A_e/A_i = 0.75$, $a = 366$ m/s, $C_i = 0.35$. Copyright © 1968 and new material © 1985 by The Massachusetts Institute of Technology.

and we cannot draw conclusions based entirely on calculations for flat cylinder heads. Two inlet valves can be used in various ways (by opening them different amounts, for example) to generate swirling (rotation with a vertical axis) or tumbling (rotation with a horizontal axis) flows in the cylinder. We will talk about such flows in Chapter 5. There are also aspects of combustion chamber design other than the ratio of valve area to bore area, and some of these can interact with the number and placement of the valves; squish, for example (see Chapter 5), and the combustion chamber shape may be determined partially by a desire to strain a tumbling flow to produce high turbulence levels, something we will also discuss in Chapter 5. Finally, it seems possible that some of the popularity of four valves per cylinder may be attributed to market forces – customers expect four valves per cylinder, and would not know what to make of three, even if the performance were superior.

The motivation for making the intake valve the singleton, and doubling the exhaust valve, is cooling: in Chapter 3 we will see that the smaller the exhaust valve, the shorter the path the heat must take to a sink, and the cooler the valve. The intake valve does not have much of a cooling problem, and hence can afford to be larger. Also, note that the exhaust valve area is 0.75 of the intake valve area. This is common practice. The exhaust valve opens some 57° BBC, when there is still considerable pressure (perhaps 3.45×10^5 Pa) in the cylinder to help get the exhaust gases moving out the valve. By contrast, the intake has only atmospheric pressure (roughly 1.0×10^5 Pa) to get the gases into the cylinder. Hence, the exhaust valve area can afford to be a little smaller.

It is clear that four valves (or three) have a substantial advantage over two, in that they will permit a higher engine speed before the flow chokes; at the same torque, greater speed means greater power. Between 1912 and 1922, most of the DOHC (double overhead cam) racing engines were built with four valves per cylinder. However, in 1923, nearly everyone changed to two valves per cylinder, and stayed that way until after the Second World War. This curious phenomenon can be attributed to the fact that Fiat in 1923 had an OHC engine which attained the then magic figure of 41.8 kW/L (56 HP/L) using only two valves per cylinder [17]. Ernst Mach (of the Mach number – 1838–1916) had done his great work, and died. However, the question of valve flow capacity and the Mach index that we have been examining does not seem to have been explored until 1942 [53], [65], in connection with the Second World War. Hence, the other designers in 1923 had no idea why Fiat had been so successful, and presumed it had something to do with the two valves per cylinder. This was counter productive.

Now, as we have seen, several normally aspirated engines (for example, the Ferrari with five valves per cylinder) manage to achieve figures in the neighborhood of 75 kW/L(=100 HP/L).

Regarding the early appearance of four valves per cylinder, as I remarked in the preface, shortly after the First World War, Frontenac marketed a double overhead cam, four valve per cylinder, chain-driven conversion for the Model T Ford, which sold in large numbers, and came to dominate dirt-track racing in the United States [17]. I include here a picture of the conversion. Multiple valves per cylinder have

Figure 2.8. Two views of the chain-driven 16-valve Model DO Frontenac conversion on a Model T Ford engine. The engine was on display in Harrah's Automobile Collection [17]. Reproduced by kind permission of the Estate of Griffith Borgeson.

been fairly common since the automobile reached its mature form in the first decades of the century.

For a fixed number of valves, fixed displacement and a flat head, the Mach index can be reduced by reducing the piston speed, that is, by reducing the S/b ratio. However, this increases the lateral dimensions of the combustion chamber, increasing the burn time, a problem that might be resolved by multiple spark plugs. In addition, while the volume of the combustion chamber is held fixed, the surface area increases, increasing the heat loss and reducing η_i.

A more effective solution to the problem of maximizing $(D/b)^2$ is a change in the shape of the combustion chamber. A domed or pent-house shape ("hemi-head" – see Figure 1.11, "penta-head" – see Figure 2.11) not only has a lower surface/volume ratio, but makes it possible to have much larger valves, although this requires more complicated valve gear, since the valves now must operate at an angle. We reproduce here (see Figure 2.9) a figure from [95] which illustrates the advantages.

Note in Figure 2.9 that Taylor is using slightly different notation from ours – he is assuming that the valve seats are slightly larger in diameter than the valve heads, and he is calling these respective diameters D_{si} and D_i for the intake valve. It is evident that, by taking the dome up to $H/b = 0.5$, a full hemisphere, the piston speed V_p at $Z = 0.5$ has been increased by a factor of 2.33 over a flat head. Referring to the Jaguar XK engine again, it had a value of $H/b = 0.4$, permitting $D_i/b = 0.536$, which is a little less than Taylor believes would be possible.

If the combustion chamber is domed, and the piston crown is flat, the clearance volume will usually be too large at modern compression ratios. It is consequently necessary to have a domed piston crown that extends into the combustion chamber to reduce the clearance volume somewhat. In Figure 2.10 we show the form of the XK pistons used in the 3.4 L engine for various compression ratios.

We will discuss in Chapter 5 the effects of squish. Briefly, the clearance volume at TC may cover only a part of the piston crown. As the piston approaches TC, the

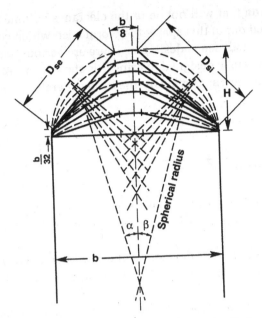

Figure 2.9. Valve-area ratios for domed cylinder heads [95].

$\frac{H}{b}$	$\frac{D_i}{b}$	$\frac{A_i}{A_p}$	$\alpha°$	$\beta°$	V_p at $Z=0.5$ m/sec (ft/min)
0.50	0.65	0.42	49.3	43.9	25.1 (4980)
0.44	0.61	0.37	45.2	40.6	22.2 (4400)
0.38	0.56	0.31	40.4	36.6	18.5 (3680)
0.31	0.53	0.28	35.1	31.9	16.7 (3320)
0.25	0.48	0.23	29.1	26.6	13.8 (2730)
0.13	0.44	0.19	15.7	14.1	11.4 (2260)
0	0.42	0.18	0	0	10.8 (2140)

$A = \pi D^2/4$, where D = valve diameter, D_{si} = inlet valve seat diameter, D_{se} = exhaust valve seat diameter. It is assumed that $A_e/A_i = 0.81$. The values of V_p are determined at $D_i = 0.975 D_{si}$, $D_e = 0.975 D_{se}$, $a = 366$ m/s, $C_i = 0.33$ and $Z = (\frac{b}{D_i})^2 \frac{V_p}{aC_i}$. Copyright © 1968 and new material © 1985 by The Massachusetts Institute of Technology.

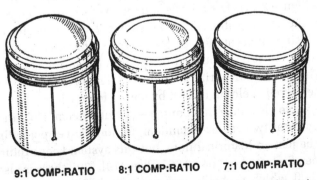

9:1 COMP:RATIO **8:1 COMP:RATIO** **7:1 COMP:RATIO**

Figure 2.10. Jaguar 3.4 L pistons [2]. Reproduced by kind permission of Jaguar Cars, England.

space above the piston that will not be in the clearance volume disappears, and as the gases are forced out of this space, they form a jet, which contributes to the turbulence, which we will see is desirable. There are various ways in which the combustion chamber can be designed to bring this about, some of them involving dishing of the piston crown. Obviously these considerations interact with our considerations above.

2.9
VALVE ACTUATION

At the present time nearly all four-stroke engines use poppet valves opened by a cam and closed by a spring. We will look at some of the problems with this arrangement in a moment.

At one time or another in the development of the internal combustion engine, several different types of valve and valve actuation mechanism have been tried. Most of these have disappeared without a trace. The nearest competitor of the poppet valve and spring was probably the sleeve valve, which had a long run of popularity during the twenties. In this system, the piston ran in an inner sleeve, which ran in an outer sleeve, which ran in the cylinder bore. Each sleeve had ports. The sleeves were pushed up and down by pushrods operated by a cam, bringing the ports into alignment at the right time, to allow the gases to enter the cylinder and to escape. The system was extremely quiet (one of the most popular cars was the Silent Knight) and trouble-free, but it required that the sleeves be flooded with oil. Some of this oil inevitably found its way into the cylinder, so that these cars were followed by a cloud of blue smoke wherever they went. This was probably responsible for the eventual demise of this system, although it was surely helped by the introduction of other methods of quieting tappet noise – the hydraulic valve lifter, but also valve covers with double walls containing sound-absorbing material (The Armstrong Siddeley, for example).

Another system that is still with us from time to time is the desmodromic valve. The name means "resembling a racetrack," and refers presumably to the stirrup used in the first example on the 1914 Delage (see Figure 2.11). A desmodromic system can take many mechanical forms. In the 1914 Delage system, there are three cams (see Figure 2.11). The middle cam opens the valve in the normal way, pressing against the opening shoe; on either side of it are the closing cams, which act against the closing rollers at the top of the stirrup. The valve is captive between the two cams, and is positively pushed in and out. No spring is required to hold it on the cam. There is a small spring in the valve stem to allow it to extend somewhat, so that the valve is held on its seat by spring force when closed – this takes the place of the valve clearance in a poppet – spring system. Note in Figure 2.11 that each cam operates two valves in tandem, for four valves per cylinder.

This would be just an historical note, but this system keeps returning. It appears to have been first used on the 1914 Delage [19], [20]; there is also a drawing of a 1914 Fiat [18] which probably was never built (because of the beginning of the First World War). It was again used on the 1922 Rolland-Pilain, and then

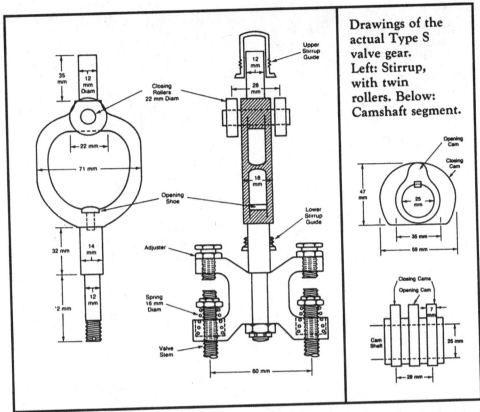

Drawings of the actual Type S valve gear. Left: Stirrup, with twin rollers. Below: Camshaft segment.

Figure 2.11. The desmodromic valve system as used on the 1914 Delage [19], [20]. Reproduced by kind permission of Jack Nelson, Romsey, Victoria (Australia). Sketch by Nelson, redrawn by Rebecca Stuckey.

reappeared on the 1954 Mercedes-Benz, and again on the 1991 Mercedes-Benz. In fact, it has been used continuously from 1955 to the present by the Ducati motorcycle, which has been very successful in racing. It will probably make another unexpected appearance, since it can substantially improve gas mileage (see Problem 1 in Chapter 6).

To understand the reason for the resurgence of this system, we must understand the problems with the poppet valve – spring system. The spring is required to hold the valve on the cam during the downward part of the valve motion. The spring must be strong enough to hold the valve train on the cam at the maximum engine speed, when accelerations reach values of several hundred g. That is, the spring force must equal the mass of the valve train multiplied by the maximum acceleration of the valve train. When the cam is opening the valve, it must not only accelerate the valve upward, but it must do it against the force of the spring. At maximum engine speed, this increases the load on the cam by roughly 50%. At low engine speeds, however, when the acceleration of the valve train is negligible, the cam must still push against the spring, a load equal to one-half the inertial load at maximum acceleration. We will analyze these forces in a moment. This loading results in fatigue problems on the surface of the cam follower, and in

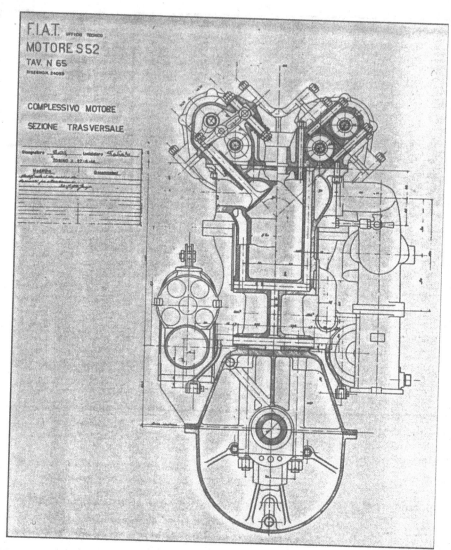

Figure 2.12. Transverse section of the FIAT S52 motor of 1914 [18] with desmodromic valve gear. Note that this is an excellent example of a penta-head. Reproduced by kind permission of Fiat Auto.

severe lubrication problems, as well as causing the cam to contribute 25% of the engine friction (the pistons and rings, cam, bearings and oil pump consume nearly 20% of the power, so the cam is responsible for roughly 5% of the power). In addition, when the engine speed exceeds the speed at which the spring force is equal to the inertial force, the cam follower no longer stays on the cam, and the valve opens farther, in free-fall, coming down to bounce on the cam. This is known as valve float, or valve bounce.

The desmodromic valve system eliminates the spring force. Now the cam is only required to supply the force to accelerate the valve train, which is much lower. This vastly reduces the engine friction at low speeds, and permits much

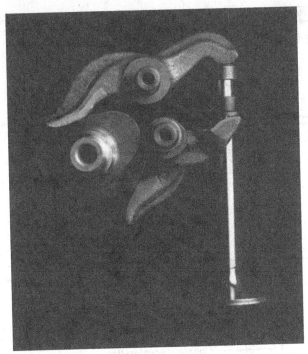

Figure 2.13. The design of the Ducati desmodromic valve system. Reproduced by kind permission of Ducati North America Inc.

higher engine speeds. Valve float is also largely eliminated. As engine speed, and inertial loading of the valve train, increases, the desmodromic mechanism will no longer be approximately rigid, and this will permit a sort of float, but probably of smaller magnitude. This is at the cost of somewhat greater complexity and manufacturing cost, and possibly lower reliability.

Let us now give our attention to the poppet-spring system. In Figure 2.14 we show typical (for a production vehicle) curves of lift, velocity and acceleration of a poppet valve held on its cam by the spring. The constant velocity ramp at each end is required to close the valve clearance. If hydraulic tappets are used (not ordinarily in performance engines), this ramp takes up the slack in the tappet, and the ramp can be steeper. The impact stresses when the clearance closes are a function of the ramp velocity. Taylor [95] quotes a figure of 0.63 m/s for large aircraft engines.

The remainder of the curve consists of a period of uniform upward acceleration produced by the cam, a period of nearly uniform downward acceleration (determined by the cam profile, to which the valve train is held by the spring), and another period of uniform upward acceleration provided by the cam.

The lift curve (including the ramps) is fairly closely approximated by $(L/2)(1 - \cos \omega t)$, and this is the default form that is used by ESP. However, the velocity and acceleration produced by this curve are considerably in error – small differences in slope and curvature that are not particularly apparent to the eye become

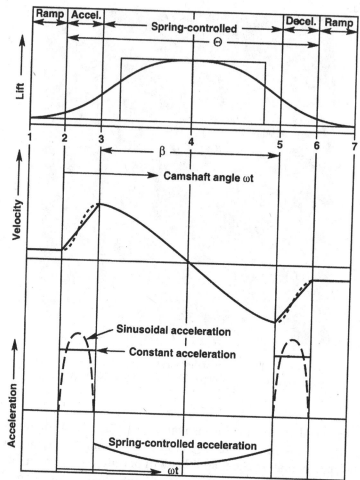

Figure 2.14. Theoretical valve lift, velocity and acceleration vs. camshaft angle. ω = crankshaft angular velocity, t = time [95]. Copyright © 1968 and new material © 1985 by The Massachusetts Institute of Technology.

very important when a curve is differentiated once or twice. Since ESP is only interested in the lift, this is not important.

Ordinarily, the valve lift is defined as the lift above the ramp, and the valve opening angle as the angle between points 2 and 6 on Figure 2.14. For the XK engine, for example, if it followed the pattern of Figure 2.14, the ramp would amount to 32 CAD, the upward acceleration to 35 CAD, and the opening time to 240 CAD. The ratio of the time between 2 and 3 to the time between 2 and 6 is about 1/7 on Figure 2.14. The cosine curve rises rather more than the ramp does; for a valve with a lift of 11 mm, the lift at the end of the ramp using the cosine curve is about 1 mm, while it should realistically be about 0.3 mm. However, there is not much flow through either gap, so it does not make much difference to ESP. However, it is necessary when using ESP to offset the opening and closing times to allow for the ramp if using the cosine program.

We can do a simple analysis of the forces involved if we approximate the spring-controlled acceleration as uniform (so that the middle section of the velocity curve is a straight line), and ignore the ramp velocity. Suppose that the total time between the point 2 and the point 6 is T (that is, the time exclusive of the ramps), and that the time between 2 and 3 is Δ. Let the upward acceleration between 2 and 3 (and 5 and 6) be a_0. Then, using the condition that the curve must be symmetric, and the velocity and position of the valve must return to zero, we find that the maximum lift

$$L = a_0 \Delta T/4 \qquad\qquad (2.20)$$

If $\Delta = T/7$, we have $L = a_0 T^2/28$. Again using the Jaguar XK 3.4 L engine as an example, at 5500 rpm, $T = (1/2N)(240/180) = 7.27 \times 10^{-3}$ s. The lift is $L = 9.53$ mm. We find an acceleration $a_0 = 5.05 \times 10^3$ m/s^2 = 515 g. This is the upward acceleration; the downward acceleration provided by the spring is only 2/5 of this (if $\Delta/T = 1/7$), or 206 g. If we approximate the mass of the valve train (valve, keeper, tappet, one-half the mass of the spring – since one end is stationary and the other end moves with the valve) as 0.2 kg, the spring force is approximately 4×10^2 N ($= 90$ lb$_f$). We are assuming that the spring force just produces the observed acceleration at this engine speed – that is, that the valve will float if the engine goes any faster. At lower engine speeds, a fraction of the spring force (larger as the speed is lower) is devoted to pressing the tappet against the cam, since it is not needed to produce the lower acceleration of the valve train at these speeds. The specification for the Jaguar XK valve springs as fitted (that is, at the length corresponding to the closed valve) is 3.5×10^2 N ($= 78.7$ lb$_f$). The spring force at maximum lift is 5.95×10^2 N ($= 134$ lb$_f$). My estimate is just a little under the average of the two, which suggests either that I slightly underestimated the mass of the valve train, or that the valves do not float until slightly above the engine speed for maximum horsepower, possibly both. Incidentally, the mass of the valve train in a pushrod engine is probably nearly double that in an overhead cam engine, which requires stiffer valve springs, greater losses, lower speed at valve float and so forth.

Note that, in general the spring acceleration a_1, say, is

$$a_1 = \frac{a_0}{\frac{T}{2\Delta} - 1} \qquad\qquad (2.21)$$

which simply states that the integral of the acceleration over the open period must vanish. Hence, as Δ is reduced, a_1 is reduced, but a_0 is increased. Getting the valve to open earlier, and close later, which would increase the flow coefficient at the shoulders of the curve, would also increase the load on the cam follower. When a high lift cam is installed, it is often accompanied by roller followers, to alleviate the fatigue and wear associated with the higher loading. A roller follower has a roller bearing in contact with the cam lobe.

We include here Figure 2.15, which shows valve bounce, and how it can be controlled by a stronger spring. Note that the speed is the cam rpm, which is half the crank rpm, and the degrees are cam degrees and not crank degrees.

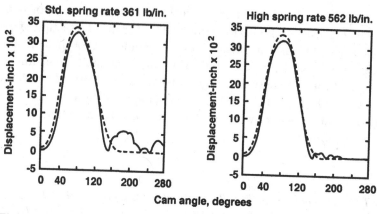

Figure 2.15. Valve performance with two different cam follower springs. GM series 53 engine; cam speed 4,000 rpm [52]. Reprinted with permission from 620289 © 1962 Society of Automotive Engineers, Inc.

2.10

VALVE TIMING

The gases that are flowing into and out of the engine have inertia; once they are moving, they want to keep moving. Figure 2.16 is an example of valve timing for a passenger vehicle, presented in a traditional way.

The exhaust is opened 57 CAD BBC. This makes use of the pressure in the cylinder to get the exhaust gases moving out of the cylinder. The exhaust valve is not closed until after TC, and the inlet valve is opened before TC. For 30 CAD both valves are open together, though only about one-half open at worst. This is called overlap. At WOT, the exhaust gases rushing out of the exhaust valve help to pull fresh charge into the cylinder, getting the gases in the manifold moving even before the piston has moved appreciably. (At partial throttle, when the pressure in the inlet manifold is below that in the exhaust manifold, the situation is much more complicated, involving backflow of exhaust gases into the inlet manifold.) The inlet valve is held open 57 CAD ABC, although it is only about 60% open at BC, because the fresh charge, rushing into the cylinder, does not want to stop, and continues to rush in even after the

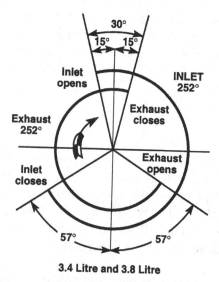

3.4 Litre and 3.8 Litre

Figure 2.16. Valve timing for the Jaguar 3.4 L and 3.8 L engines [2]. Reproduced by kind permission of Jaguar Cars, England.

piston has started up. The inlet valve is not closed until the gases are brought to a halt by the rise in pressure in the cylinder, which maximizes the mass packed into the cylinder.

This effect will be different, depending on the length and configuration of the inlet manifold and of the exhaust manifold, since these are the gas columns that are being accelerated and decelerated. The effect will also change with engine speed. Something that works at high speed will not work at low speed.

The gas in the cylinder is, of course, compressible, and acts as a spring; the mass of gas in the manifold works against this spring, forming an equivalent spring-mass system. This system has its own characteristic frequency. We will analyze the characteristics of this system in Section 2.12.2. For street vehicles, the frequency associated with this spring-mass system is usually too high to match the frequency of engine rotation. The losses to the turbulent boundary layers in the manifold are absolutely negligible – the effect is of order one part in 10^6 compared to the inertia. However, when the valve is partially open, the pressure drop associated with the acceleration of the manifold flow through the relatively narrow gap is quite comparable to pressures generated by inertia, and acts to slow down the oscillation substantially, making the frequency of the sloshing gas and of the engine rotation comparable. We will examine this in greater detail, making use of ESP, in Section 2.12.2.

As we have seen, the brake power of the engine peaks and then falls off, for a number of reasons: principally the fall in volumetric efficiency, and the fall in mechanical efficiency. We cannot improve the volumetric efficiency by opening the valve wider, as we have seen. Supposing that we cannot change the diameter or number of the valves, one thing we can do is keep the valves open longer; the pressure in the cylinder is lower than it could be, and if the valve is open longer, more mass will enter the cylinder. This will only work if it takes place at a higher engine speed, making use of inertia. In Figure 2.17 we show the power P_b vs. engine speed in rpm for a Chevrolet small block engine in production form, with a mild race cam (the kind that is sold for street use), and in NASCAR form (NASCAR = National Association of Stock Car Automobile Racing). The NASCAR engine has not only what is called a full race cam, but a number of other modifications in addition.

If we look at the curve of valve opening in Figure 2.14, it is evident that there is a considerable period at each end during which the valve opening is small. We may make a crude estimate of the effective open period by comparing the true curve with a fictitious square curve in which the valve opens fully instantly, stays fully open for a period, and then snaps shut instantly. The total integrated open area must be the same in the two cases.

For the curve of Figure 2.14, we may calculate this effective open angle

$$\Theta_{eff} = \left(1 - \frac{\Delta}{T}\right) 2\Theta/3 \qquad (2.22)$$

where Θ is the true open angle, Δ is the angular width of the upward acceleration

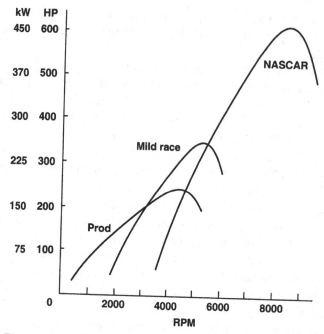

Figure 2.17. Performance of production, mild race and full race modified Chevrolet small block engines.

	Production Engine	Mild Race	NASCAR
IO CAD BTC	3	20	60
IC CAD ABC	47	50	70
EO CAD BBC	47	60	90
EC CAD ATC	3	10	40
P_b kW (HP)	179 (240)	261 (350)	447 (600)
N rpm	4500	5500	8500

pulse, and T the total open angle. We are using interchangably Δ and T as either time or angle. When $T/\Delta = 7$ as it does in Figure 2.14, this gives $\Theta_{eff} = 0.571\Theta$. We have sketched this square curve on Figure 2.14.

In the case of the production Chevrolet, the inlet open angle is 230 CAD; let us suppose that the cam profile is the same as in Figure 2.14. Then the effective open angle Θ_{eff} is 131 CAD. Hence, if we take the true open angle $\Theta = 230$ CAD, and subtract the effective open angle 131 CAD, we have 99 CAD left, and half of this can be deducted at each end, to give approximate equivalent opening angles of 46.5 CAD ATC and 2.5 CAD BBC. Bear in mind that my estimate of the effective opening and closing point is very crude – that is, the definition of equivalent, while precise, may not always be relevant. Hence, the 2.5 CAD BBC is not particularly significant. We can only say, that the effective closing point is roughly BC, which makes some sense at the speeds of the production engine. The opening point, however, seems late; why not at TC? That is, why not have the real opening point some 40 CAD BTC? The problem is, that when the engine runs at part throttle, the pressure in the inlet manifold is quite low, and if there is significant overlap between the inlet and exhaust valves, the pressure differential between the inlet and exhaust manifold will force exhaust gases into the inlet manifold. Also, at wide-open throttle, gas inertia will cause fresh charge to be lost out the exhaust.

Looking now at the mild race engine, the overlap has been increased somewhat, from 6 CAD to 30 CAD. From our discussion above, we would not expect this engine to idle very well, and it does not (see Figure 2.17). Idle, such as it is, has moved up from some 500 rpm to nearly 2,000 rpm. However, at full throttle, inertia helps to get a fuller charge into the cylinder. The total angle of opening has been increased to 250 CAD.

In determining the effective opening and closing points, we must consider whether the cam profile has remained the same shape; that is, is the ratio T/Δ still roughly equal to 7? Looking at Figure 2.17, we see that the speed at the peak power has gone up from 4,500 rpm to 5,500 rpm. It is a reasonable presumption that the power peaked at 4,500 rpm in the production version as a result of the Mach index's reaching a value of 0.6. Increasing the angle of opening while keeping T/Δ the same will not change the Mach index, since C_i will not change. If we can take $C_v \propto L$, which is approximately true up to $L/D = 0.25$, then we can write

$$\Theta_{eff} = \frac{1}{C_v^{max}} \int_{\theta_1}^{\theta_2} C_v \, d\theta \tag{2.23}$$

Combining this with Equation 2.14, we can immediately write

$$\frac{\Theta_{eff}}{\Theta} = \frac{C_i}{C_v^{max}} \tag{2.24}$$

Hence, to change the Mach index, C_i must change in proportion to the change in the peak power rpm. Thus, we expect

$$\frac{C_i}{C_v^{max}} = \frac{5500}{4500} \times 0.571 = 0.698 \tag{2.25}$$

If we look at Equation 2.22, it is evident that even if $\Delta/T = 0$, we cannot achieve a value of Θ_{eff}/Θ greater than 2/3. Hence, it is not enough to simply change the value of the ratio Δ/T. We must go to a different cam profile altogether. If we consider a profile like Figure 2.18, that is: a uniform upward acceleration a_0 from 0 to Δ, and an equal, uniform downward acceleration $-a_0$ from Δ to 2Δ, and the opposite at the other end. This gives

$$L = a_0 \Delta^2 \tag{2.26}$$

which gives much higher values of acceleration, and hence cam follower load, that the profile of Figure 2.14. This cam will certainly need roller followers. For

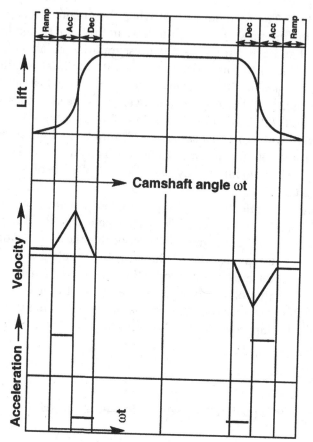

Figure 2.18. Curves of lift, velocity and acceleration for a possible high lift cam profile.

this cam profile,

$$\frac{\Theta_{eff}}{\Theta} = 1 - \frac{2\Delta}{T} \tag{2.27}$$

For a value of 0.698 (see Equation 2.25) we will need $T/\Delta = 6.62$, which is not too different from Figure 2.14, but the spring force now must produce $-a_0$ instead of $-2a_0/5$. The spring is therefore 5/2 as strong.

To return to the effective opening and closing angles: we had $\Theta = 250$ CAD, so that $\Theta_{eff} = 175$ CAD. Half the difference is 38 CAD, giving an effective opening angle of 18 CAD ATC, and a closing point of 12 CAD ABC. More use is made of inertia at both opening and closing.

The NASCAR engine has been extensively modified. More than a different cam profile is needed to produce this performance. If we make a calculation similar to Equation 2.25, we find that C_i must be larger than C_V^{max}, which is not possible. The valve open area must have been changed. The total open angle of the inlet valve is now 310 CAD. If we keep Δ the same as for the mild race configuration, so that the valve springs need not be stronger, we will have $T/\Delta = 8.2$, so that

$\Theta_{eff} = .756\Theta = 234$ CAD. Making the same calculation as before, we find that now the inlet valve is (effectively) opening at 22 CAD BTC, and (effectively) closing at about 32 CAD ABC. Probably considerable mixture is being lost out the exhaust, and we can certainly not expect this engine to idle, but neither matters in a racing engine. Considerable use is being made of inertia in determining the inlet valve closing. Both the peak power, and the engine speed when it occurs, have risen considerably. The engine will now not run below about 3,500 rpm. I have seen Formula I engines "idling," blowing a mist of raw fuel backward out of the carburettors.

2.11
VARIABLE VALVE TIMING

Clearly, there would be advantages if the cam profile of a given engine could be changed during operation, going from a production profile at low engine speeds to a mild race profile at intermediate speeds, and on to a full race profile at high speeds. Then the performance curve of the engine could follow the envelope of the different curves in Figure 2.17; the engine would idle satisfactorily, but would peak at a much higher speed.

A number of ways have been put in production to accomplish some part of this goal. The simplest, and least expensive, possibility is known as cam phasing. This is only applied to DOHC engines. One camshaft is rotated relative to the other one, changing the phase of the inlet relative to the exhaust. This does not change the shape of the cam profile, so that it cannot change the total open angle, but it can change the overlap and the closing angle of the inlet valve. In Figure 2.19

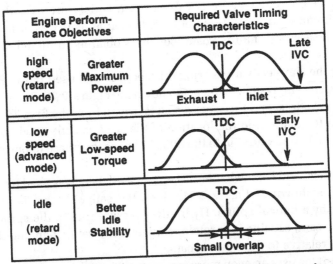

Engine Perform-ance Objectives		Required Valve Timing Characteristics
high speed (retard mode)	Greater Maximum Power	TDC, Late IVC, Exhaust, Inlet
low speed (advanced mode)	Greater Low-speed Torque	TDC, Early IVC
Idle (retard mode)	Better Idle Stability	TDC, Small Overlap

Figure 2.19. A cartoon of valve lift characteristics with cam phasing [30]. Reprinted with permission from 910447 © 1991 Society of Automotive Engineers, Inc.

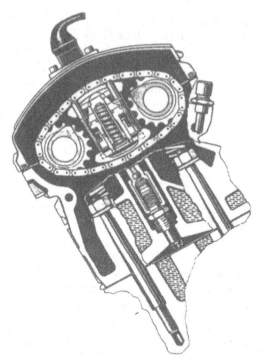

Figure 2.20. The Porsche 968 Variocam system [16]. Reprinted with permission from *Automotive Engineering* Magazine, © 1991 Society of Automotive Engineers, Inc.

we show a cartoon of valve lift with cam phasing. At low speed, the inlet cam is retarded, to reduce overlap, to give better idle. As the speed increases, the inlet cam is advanced, to close the inlet valve earlier, which improves the low-speed torque; the increase in overlap is not as detrimental as it was at low speed. As the speed increases still more, greater benefit is obtained from delaying the inlet closing.

Probably the simplest way to bring this about is that used in the Porsche 968, called the Variocam system, which is shown in Figure 2.20. Here, one cam is driven from the crank, and the second cam is driven from the first. The chain connecting them has slack, and the location of the slack is controlled hydraulically. Moving the slack from one side to the other changes the phase of one cam relative to the other.

A more direct, and somewhat more complicated, way was used by Alfa Romeo and Nissan, as shown in Figure 2.21. Here, the cam sprocket is connected to the camshaft by a helical spline. Hydraulic pressure moves the sprocket axially relative to the cam, causing it to rotate relative to the cam, changing the phase of the inlet cam relative to the exhaust cam.

Mercedes-Benz uses an even more complex system, shown in Figure 2.22. This system involves an annular piston to rotate the sprocket relative to the cam. Some

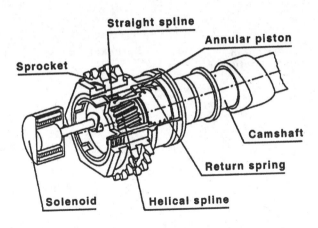

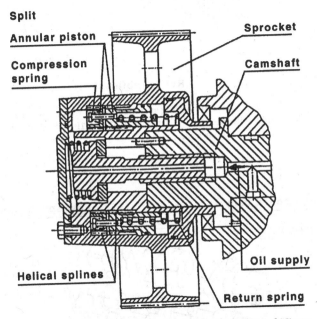

Figure 2.21. The cam phasing system used by Alfa and Nissan [30]. Reprinted with permission from 910447 © 1991 Society of Automotive Engineers, Inc.

of this mechanical ingenuity has a technical motivation, but some of it is driven by the desire to avoid paying royalties to someone else for a patented invention.

Although it is not precisely cam phasing, I will include here the system used by Honda, which has a similar motivation. This is pictured in Figure 2.23. This engine has four valves per cylinder. For the two inlet valves, there are three cam lobes, and three rocker arms. The rocker arms can be locked together by hydraulic pistons. At low speeds, the rocker arms are unlocked, and operate independently; the two outer rocker arms, running on the two outer cam lobes, operate the valves, and the lift is different for each of the valves. In fact, one is opened only a very

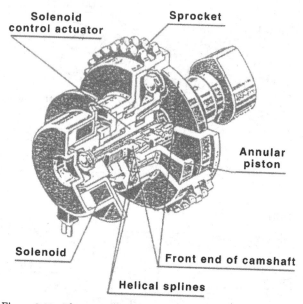

Figure 2.22. The cam phasing system used by Mercedes Benz. Reprinted with permission from 910447. © 1991 Society of Automotive Engineers, Inc.

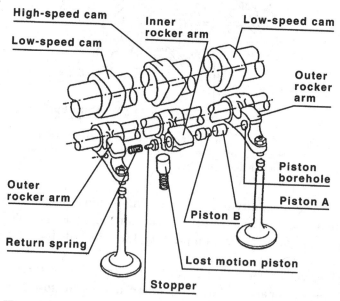

Figure 2.23. Valve gear in the Honda VTEC engine [30]. Reprinted with permission from 910447 © 1991 Society of Automotive Engineers. Inc.

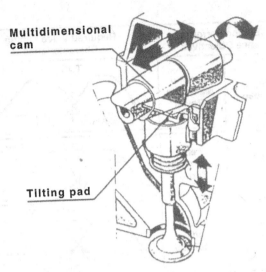

Multidimensional cam

Tilting pad

Figure 2.24. The Ferrari V8 variable valve timing system [30]. Reprinted with permission from 910447 © 1991 Society of Automotive Engineers, Inc.

small amount compared to the other, which induces a strong swirl in the incoming flow. At high speeds, the hydraulic pistons lock the rocker arms together, and the center cam lobe operates the valves. Now both valves open the same amount. In this way, Honda can have small overlap and early closing at low speeds, and large overlap and late closing at high speeds. This system has the disadvantage, however, of having to change abruptly from one to the other somewhere in midrange. In addition, the single high-speed cam experiences high loads.

Finally, the Ferrari V8 variable valve timing system is shown in Figure 2.24. This is a genuine variable valve timing system, as opposed to a cam phasing system; that is, the shape of the lift curve, and the opening and closing points, can be independently changed. Examining Figure 2.24, it can be seen that the cam lobe is rather long, and the cam profile changes from one end to the other. Since the cam lobe is no longer cylindrical, the angle made by the line of contact with the valve axis will change as the cam rotates. The cam follower has a rocking pad which follows this changing angle. To change the effective profile, the cam is moved axially by hydraulic pressure. In Figure 2.25 is shown a cartoon of the effective cam profiles generated by this system. At idle, the overlap is small, and the lift is small, with early closing. This is more efficient than throttling, since the manifold pressure is not reduced. As the speed increases, the lift is increased, as well as the duration, the inlet closing later and later. The overlap is increased somewhat, but even at high speed the overlap is not exaggerated, probably to keep loss of fuel to the exhaust under control.

Finally, I show what may be the ultimate in valve gear, the Lotus active electro-hydraulic valve train, in Figure 2.26. Here, the valves are operated hydraulically by servo-valves feeding hydraulic cylinders. Any cam profile within reason can

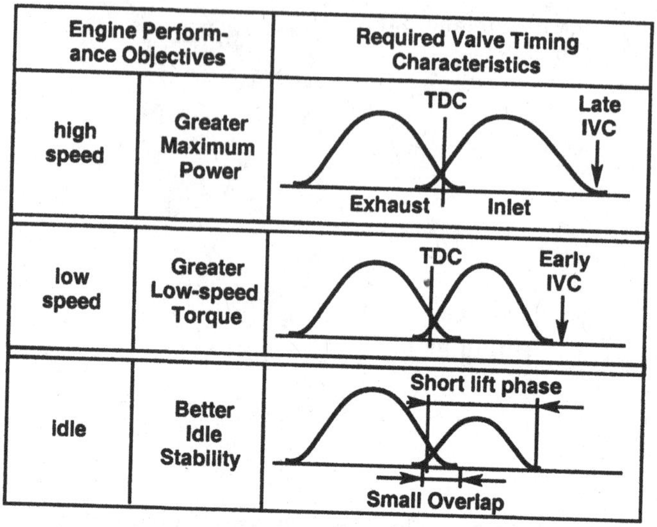

Engine Perform-ance Objectives		Required Valve Timing Characteristics
high speed	Greater Maximum Power	TDC — Late IVC / Exhaust — Inlet
low speed	Greater Low-speed Torque	TDC — Early IVC
idle	Better Idle Stability	Short lift phase / Small Overlap

Figure 2.25. Cartoon of the valve lift characteristics of the Ferrari V8 system [30]. Reprinted with permission from 910447 © 1991 Society of Automotive Engineers, Inc.

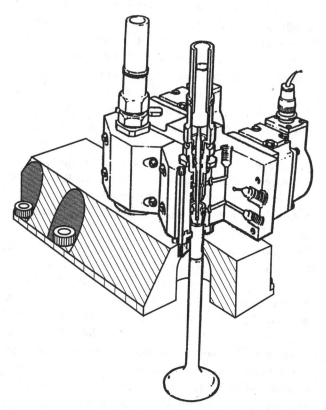

Figure 2.26. Dopson & Drake model electro-hydraulic valve train, as used on the Lotus [5]. Reproduced by kind permission.

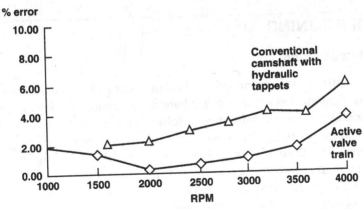

Figure 2.27. Error comparison, hydraulic/mechanical [5]. Reproduced by kind permission.

be given to the computer, and the servo system will reproduce it. In Figure 2.27 the error (the maximum relative difference in valve position) produced by a conventional camshaft with hydraulic tappets is compared with the error produced by this hydraulic system. The error is substantially lower, although both are rising toward the higher engine speeds. The hydraulic system must be designed for a maximum speed, and here the maximum has been selected as roughly the same as for a mechanical system. The response of the system has a maximum frequency for which it is designed, and this can be seen in Figure 2.28. When the valve is asked to produce a square wave form, it cannot – it has a finite rise time, and rings slightly, from which the system maximum frequency can be calculated. Designing a much higher frequency response into the system increases the cost dramatically. The cost of this system is already high, since each servo valve costs roughly $1,000 U.S., and one is required for each valve.

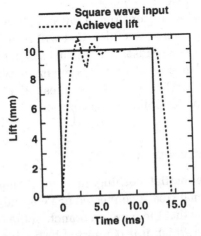

Figure 2.28. Response of hydraulic system to square wave input [5]. Reproduced by kind permission.

2.12

MANIFOLD TUNING

2.12.1 Introduction

The intake manifold, valves and cylinders form a complex system somewhat like a collection of connected organ pipes. There are resonant cavities formed by the cylinders with open valves, and by manifold branches leading to closed valves; there are changes in impedance, when a pipe experiences a change in cross-sectional area. The fluid in this system (air) is sloshing back and forth due to its inertia, bouncing against the resiliance of the compressed gas in the resonant cavities; there are compression and expansion waves traveling through the gas, reflecting from closed and open ends, and from changes in cross-section. It is clear that this is a very complicated system. It has been formally attacked by the methods of computational fluid dynamics by the faculty of Imperial College of the University of London. That is, they have programmed and solved the compressible Navier Stokes equations (which describe the motion of a compressible fluid) in a realistic geometry. The program they have produced is in use by several automobile companies.

Such a calculation is surely the proper approach to this situation, if what is wanted is hard numbers to use in serious design. However, such a program does not contribute much to our physical understanding of the situation. Such understanding comes only from building simple physical and mathematical models that behave more or less like the real situation. It is only when we have stripped away the unimportant details, leaving a situation simple enough for us to comprehend the mechanisms involved, a situation which yet displays the important behavior of the original system, that we can say we understand what is going on. Any engineering undertaking has two goals: first, we must understand what is going on, and second, we must be able to calculate to adequate accuracy. We should never try to do the second before doing the first.

Fortunately, it is possible to construct a few simple models of this situation, applying some simple physical principles, that will show us what is happening. We will then be able to use these ideas to construct a better model that can be used in ESP. This model will give results quite comparable with reality in a limited range of conditions (e.g., for four cylinder engines), and will yet be enormously simpler than the full Imperial College CFD approach.

2.12.2 Helmholtz Resonators

Let us deal first with the inertial sloshing. Imagine a simple situation, a pipe of length L and cross-sectional area A connected to a volume V (see Figure 2.29). To fix ideas, you can imagine that the pipe is a branch of the inlet manifold, and the volume is the cylinder to which it is connected by a wide-open valve. Now, the volume will act as a spring. If we push more gas into the volume, it will compress the gas already there, and raise the pressure, which will try to expel the gas we have pushed in. Suppose that the gas column in the pipe is pushed down by a

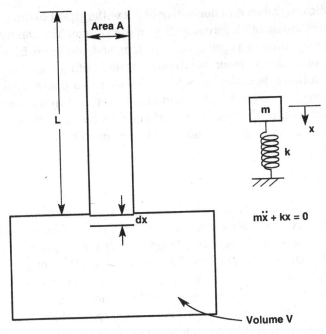

Figure 2.29. Schematic of simple Helmholtz resonator.

distance Δx (see Figure 2.29). This causes a mass

$$\Delta m = \rho A \Delta x \tag{2.28}$$

to enter the volume. The density in the volume will increase as a result, giving $\Delta \rho = \Delta m / V$. This increase in density will cause an increase in pressure. The change in pressure is related to the change in density by the isentropic gas law, since we presume that these changes are taking place reversibly and fairly rapidly (in a few milliseconds), too fast for heat conduction to be important:

$$\frac{\Delta p}{\Delta \rho} = a^2 \tag{2.29}$$

where a is the isentropic speed of sound. Now, the change in pressure times the area is the change in force:

$$\Delta F = \Delta p A = a^2 A \Delta \rho = \frac{a^2 A^2 \rho \Delta x}{V} \tag{2.30}$$

The coefficient of Δx on the right hand side is a spring constant k, of dimensions [force per unit displacement]:

$$k = \frac{a^2 A^2 \rho}{V} \tag{2.31}$$

The gas in the pipe has a mass, $m = \rho A L$, and the equation of motion of this mass is given by

$$m \Delta \ddot{x} + k \Delta x = 0 \tag{2.32}$$

where $\Delta \ddot{x}$ indicates the second derivative of Δx with respect to time. We are supposing that there are no other forces acting on the system. You can recognize this equation as the equation of a spring-mass system, and you know from experience that this is an undamped, linear, oscillatory system with a single degree of freedom, Δx. The solution is of the form $\Delta x \propto \cos \omega t$ with a phase angle depending on the initial conditions (that is, it could be proportional to the $\sin \omega t$ also – or to some combination of the two – the difference is just a shift in phase, and is irrelevant to our concerns here). The frequency is given by

$$\omega = \sqrt{\frac{k}{m}} = \sqrt{\frac{a^2 A}{VL}} \tag{2.33}$$

Now, suppose we have a $4L$, six-cylinder engine, with intake valves of diameter 45 mm. Presumably, the cross sectional area of the manifold will be the same as the maximum open area of the valve. Suppose that the manifold is 0.3 m long. Take the speed of sound $a = 343$ m/s. Take the volume as that of one of the cylinders, with the piston at BC, so that $V = 0.67 \times 10^{-3}$ m^3. $A = 1.52 \times 10^{-3}$ m^2. This gives $\omega = 9.43 \times 10^2$ rad/s, and the corresponding linear frequency is $f = \omega/2\pi = 1.5 \times 10^2$ Hz. This frequency corresponds to 9,000 rpm, which is too high for a match. Put in more practical terms, the gas velocity is a maximum about halfway through the intake stroke; it will go to zero (and start to reverse) a time later when $\omega t = \pi/2$. The entire intake stroke takes roughly $1/2N$. For a match, we want half of this to be $t = \pi/2\omega = 1/4N$, which is the same as $N = \omega/2\pi$.

The situation is a little more complicated than that, however. There is friction due to turbulent boundary layers in the pipe, but it is not difficult to show that that is a very small quantity, of the order of one part in 10^3, and hence does not play a role here.

However, the valve is only fully open for a short time, and the partially open valve results in quite a pressure drop. It is satisfactory to model this sloshing flow as basically incompressible, except for the compression of the fluid in the cylinder. Let's model the valve by an orifice of area $A_1 < A$ between the pipe and the volume. We can apply Bernoulli's equation [31] between the flow in the pipe and that through the orifice. Let $\Delta \dot{x}$ be the velocity in the pipe, and $\Delta \dot{x} A/A_1$ be the velocity through the orifice. If p_1 is the pressure in the volume, then

$$p - p_1 = \frac{1}{2}\rho \Delta \dot{x}^2 \left[\left(\frac{A}{A_1}\right)^2 - 1 \right] \tag{2.34}$$

Using Newton's second law, the equation becomes

$$\Delta \ddot{x} + \frac{\Delta \dot{x}^2}{2L}\left[\left(\frac{A}{A_1}\right)^2 - 1 \right] + \frac{a^2 A}{LV}\Delta x = 0 \tag{2.35}$$

The sign is correct, because it is $p_a - p$ (where p_a is atmospheric pressure) that drives the flow in the manifold in the positive x direction, and $p_1 = a^2 A\rho\Delta x/V$.

This equation is now non-linear, which makes it difficult to draw simple conclusions. We could linearize, however, and get some idea of what is going on.

The flow in the manifold consists of a mean value $\langle \Delta \dot{x} \rangle$ plus a fluctuation, say $\Delta \dot{x}'$: $\Delta \dot{x} = \langle \Delta \dot{x} \rangle + \Delta \dot{x}'$. If we square this, we obtain $\Delta \dot{x}^2 = \langle \Delta \dot{x} \rangle^2 + 2 \langle \Delta \dot{x} \rangle \Delta \dot{x}' + \Delta \dot{x}'^2 \approx \langle \Delta \dot{x} \rangle^2 + 2 \langle \Delta \dot{x} \rangle \Delta \dot{x}'$. We are concerned only with the fluctuations in this analysis of sloshing; the term $\langle \Delta \dot{x} \rangle^2$ will contribute only to the mean pressure difference, so we can drop it. Thus, we will replace $\Delta \dot{x}^2$ by $2 \langle \Delta \dot{x} \rangle \Delta \dot{x}$, where we have dropped the prime on the $\Delta \dot{x}$; the equation becomes linear. Now the pressure drop across the valve acts (within this linearized approximation) as a linear damping, which slows down the flow. The circular frequency now becomes

$$\sqrt{\omega^2 - \frac{\beta^2}{4}} \tag{2.36}$$

where β is the damping factor,

$$\beta = \frac{\langle \Delta \dot{x} \rangle}{L} \left[\left(\frac{A}{A_1} \right)^2 - 1 \right] \tag{2.37}$$

and when the valve is only partially opened, the frequency is very much reduced. For example, if $A_1 = 0.4\,A$, then ω is cut in half, from 9,000 rpm to 4,500 rpm. This degree of restriction is typical of production valves. This only gives us a qualitative idea of what is happening (something we could have gotten without equations) – the pressure drop through the orifice slows down the flow, and increases the time it takes it to drop from its maximum speed to zero. Now it is possible to get a better match between the engine speed and the sloshing.

We need the Helmholtz resonator to understand another aspect of the problem: after passing through the throttle, the manifold branches, with one branch going to each cylinder. In a multi-cylinder engine, the valves will be at various phases in the various cylinders. In a four-cylinder engine, at any given time the intake valve in one cylinder will be open, and the valves in the other cylinders will be more-or-less closed. With more than four cylinders, the situation becomes more complicated, since some of the other cylinders will have inlet valves that are substantially open. Let us restrict ourselves to a four-cylinder engine, which we can idealize as having one inlet valve that is open, and all the others are closed.

The branches of the inlet manifold that lead to closed valves constitute a Helmholtz resonator. We may model the inlet manifold as a pipe of length L_1 from the atmosphere to the branching point, and another of length L_2 from the branching point to the valve, with a Helmholtz resonator at the junction between the two, the branching point, of volume equal to the volume of the three manifold branches leading to closed valves. In Figure 2.30 we sketch the real manifold, and our simple model. The cylinder is shown dashed, because ESP deals with it exactly, and it does not have to be modeled as a Helmholtz resonator. We have labeled the displacement of the fluid in the three branches x, y and z; we are dropping the Δ for simplicity. We have also indicated the mechanical spring-mass equivalent of the manifold system. Here x and y are the displacement of the two masses. The variable z does not have a counterpart, but continuity requires that the sum of the displacements in the three branches entering the intersection vanish (presuming

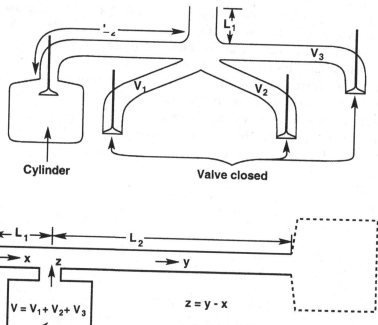

Figure 2.30. Sketch of a manifold, and of a simple Helmholtz resonator model.

that the flow is incompressible in this context), so that $z = y - x$, which is the compression of the spring. By taking pressure differences between the ends of the two pipes, it is straightforward to derive the equations:

$$\rho A L_1 \ddot{x} + \frac{A^2 \rho a^2}{V}(x - y) = 0 \qquad (2.38)$$

$$\rho A L_2 \ddot{y} + \frac{A^2 \rho a^2}{V}(y - x) = -pA \qquad (2.39)$$

where p is the pressure at the valve port. This is the correct pressure and area, since we want the force on the right-hand end of the slug of fluid in the manifold. On Figure 2.30 I have indicated the equations corresponding to the mechanical model, and it is evident that they are the same as Equations 2.38 and 2.39.

2.12.3 Organ Pipes

We have taken into account one aspect of compressibility, the compressibility of the gas in the cylinder, or of that in the dead branches of the inlet manifold. There is another aspect which we have ignored up to this point: the gas in the active branches is not just a mass that must be moved, but it is also compressible – it is like a string of tiny masses connected by tiny springs, so that waves can travel back and forth along the active branch.

We can treat these waves by using one-dimensional wave propagation theory, essentially wave-guide theory or one-dimensional acoustics [59]. In this theory, waves travel down the active pipes without change in form (so long as the pipes do not change cross-sectional area). Any change in area corresponds to a change in acoustic impedance, and results in some of the energy being reflected and some transmitted. The rules governing this theory are fairly simple, and are well-known [59]. Fortunately, we do not need to know very much of acoustic theory in order to understand what is happening here.

We need to understand that there are two types of waves, a compression wave and an expansion wave. A compression wave corresponds to an increase in pressure, and an expansion wave to a decrease in pressure.

When a wave comes to a change in cross-section, part of the wave is reflected, and part is transmitted. The transmitted wave is always of the same type as the incident wave (compression or expansion). If the incident wave sees a larger cross section, the reflected wave is of the other type (compression is reflected as expansion, and expansion as compression), and if the incident wave sees a smaller cross-section, the reflected wave is of the same type (compression is reflected as compression, and expansion is reflected as expansion). The wave consists of a disturbance of the fluid velocity and of the pressure (and density) which propagates at the isentropic speed of sound relative to the flow. That is, the wave form propagates at the speed of sound, but the velocity disturbance is much smaller. We will refer to the velocity disturbance of the wave as its velocity or intensity, hoping that there will be no confusion with the speed of *propagation*, which is the speed of sound. If u_i is the velocity of the incident wave (that is, the velocity disturbance), u_r the velocity of the reflected wave and u_t the velocity of the transmitted wave, and if S_1 is the cross-sectional area before the change in cross-section, and S_2 the cross-sectional area after the change in cross-section, then

$$\frac{u_r}{u_i} = \frac{S_2 - S_1}{S_2 + S_1} \qquad (2.40)$$

Thus, if an expansion wave of intensity 1.0 comes to a 1:4 increase in cross-section, the reflected wave is a compression wave of intensity 0.6, and the transmitted wave is an expansion wave of intensity 0.8 (so that the sound power reflection and transmission coefficients add to unity).

Let us discuss reflection of a wave at closed and open ends first, since that is the simplest case. At a closed end (an extreme reduction in area), an expansion wave reflects as an expansion wave, and a compression wave reflects as a compression wave. At an open end, however, an expansion wave reflects as a compression wave, and a compression wave reflects as an expansion wave.

Before we discuss more complicated situations, let us consider the case of a single cylinder with a single pipe for an inlet manifold. In Figure 2.31 we have sketched the situation. Shortly after the inlet valve opens, when the piston has started to drop, the pressure in the cylinder drops relative to that in the manifold,

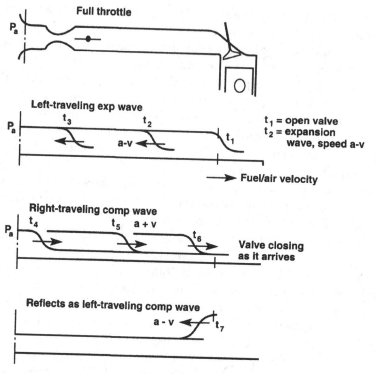

Figure 2.31. Sketch of waves in the inlet manifold of a single cylinder engine.

and an expansion wave (a reduction in pressure) starts down the inlet manifold toward the atmospheric end. It travels at a speed of $a - v$, where a is the isentropic speed of sound, and v is the fluid velocity in the manifold. This fluid velocity is changing constantly, as the piston speed changes during the cycle, but we will take it constant for the moment, for simplicity. We have seen that in a well-designed engine the Mach index should just reach 0.6 at the upper end of the performance range, suggesting that the Mach number is reaching unity at some point during the cycle. Because the cross-sectional area of the manifold branch is only slightly larger than the open area of a fully open valve, we may expect that the flow will be near sonic velocity in the manifold at some point during the cycle. Hence, we cannot neglect the fluid velocity in calculating the time it takes for the expansion wave to reach the atmosphere.

On reaching the open end of the pipe the expansion wave is reflected as a compression wave, and starts down the pipe toward the valve. Now it travels at the speed $a + v$, presumably a different v from the outward bound trip, since time has passed, but again for simplicity here we will take it to be the same. If we have arranged everything correctly the valve should be closing when the wave arrives. This means that the end of the pipe will appear to be a closed end (approximately), and the compression wave will reflect as a compression wave. A compression wave will result in an increase in pressure and this increase in pressure will help to fill the cylinder just before the valve closes. If the pipe length is L, the total transit time is

$$\Delta t_1 + \Delta t_2 = \frac{L}{a-v} + \frac{L}{a+v} = \frac{2aL}{a^2 - v^2} \tag{2.41}$$

We want this time to be comparable to the open time of the valve, say $\kappa/2N$, where κ is the ratio of the open angle to π. The condition for this to be true is

$$L = \frac{\kappa(a^2 - v^2)}{4aN} \tag{2.42}$$

If we take $\kappa = 1.3$, and $v \approx 0$, $N = 100$ rps, (with $a = 343$ m/s) then we find that $L = 1.1$ m. If $v = a/2$, $L = 0.84$ m, and if $v = 0.8a$, then $L = 0.4$ m. The velocity v is changing during the calculation, and to do this right we should integrate v during the outgoing and returning trips.

This simple model does show us that, for lower N, L is longer. Hence, we can only tune for a single-engine speed, and the effect will decrease on either side of that speed.

Now, what can we do about the existence of branches and other cylinders in a real engine? The expansion wave will still start down the active manifold branch from the valve. Let us suppose we are dealing with a four-cylinder engine. When the expansion wave reaches the branch point in the manifold, it sees a four-fold increase in cross-sectional area: three inactive branches and the air intake, which is about the same cross-sectional area as the manifold branches. This is the situation we dealt with above, so we know that the expansion wave will be reflected as a compression wave of intensity 0.6, and transmitted as an expansion wave of intensity 0.8. The wave in the air intake will be reflected from the open end as a compression wave, while the expansion waves in the inactive branches will go to the closed ends, where they will be reflected as expansion waves; when they return to the branch point, each branch will see a 4:1 increase in cross-section again; in addition, the wave that went up to the air intake, when it comes back will see a 4:1 increase in cross-section. This will result in more reflections and transmissions.

In Figure 2.32 I have sketched what is referred to as an $x - t$ diagram, which tracks the various waves as they bounce back and forth. It is clear that the wave we are interested in is the first to return, the compression wave of intensity 0.6. The next to arrive is also interesting, since it, too, is a compression wave. I have drawn the neck leading to the air intake as relatively short, so that the arrival of this wave will not be much delayed after the first; however, the air intake can be quite long and possibly folded (see Section 2.13), so that the arrival of this wave could be quite delayed. In principle, we want to arrange things, so that the first wave to return arrives at the inlet valve just as it is closing. If we do that, the arrival of the other waves is irrelevant, since the valve will be closed when they arrive. This means, that we need concern ourselves only with the distance to the branch point of the manifold; the other distances do not matter. However, if the distances are too short for this to be a reasonable time, we may make the air intake quite long, and arrange things so that the second (compression wave) to return arrives just as the valve is closing. Then we do have to concern

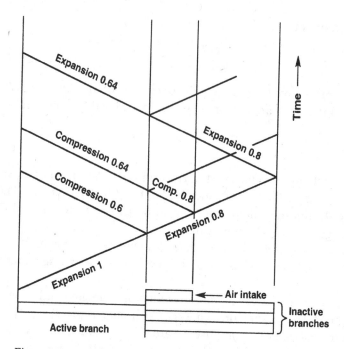

Figure 2.32. $x-t$ diagram of waves in the intake manifold of a four-cylinder engine. The nature and intensity of the waves is indicated on each ray. Only the first three arrivals at the valve are considered. Note that the flow velocity is neglected in this figure. Inclusion of the mean flow speed would change the angle of the lines, and make them not quite straight as the speed varied.

ourselves with the other waves, since they arrive when the valve is open, although they will not affect the flow into the cylinder, for reasons that I will explain in a moment.

Finally, I am including Figure 2.33 showing the tuning of the intake manifold of a D type Jaguar engine. This engine has a separate induction pipe to each cylinder. The various curves correspond to different lengths of pipe.

Note that the very long pipes cause a serious falling off of the efficiency at high rpm. A pipe in the neighborhood of 0.43m($= 17$ in) is a good compromise for this engine, and was the length usually used in racing [24].

If we are interested in making the air intake rather long, and tuning for the return of the wave reflected from this point, the situation becomes rather complicated. I have sketched this in Figure 2.34.

In Figure 2.34 we have indicated only the waves that arrive when the inlet valve of cylinder 3 is open (and whatever else is necessary in order to generate these waves). This includes waves due to the opening of the inlet valve in cylinder 1. Waves that arrive when the inlet valve is open will have a negligible effect on the filling of the cylinder, since the open valve will appear as an open end, and the boundary condition at an open end is one of constant pressure (that is what results in the type of the wave being changed on reflection, from compression to

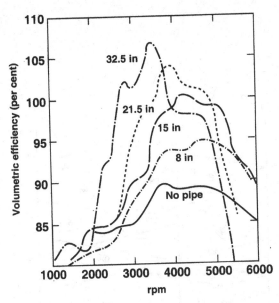

Figure 2.33. Effect of intake pipe length on volumetric efficiency (D Type Jaguar engine) [24]. Courtesy of Bentley Automotive Publishers.

expansion, and vice versa). These waves that arrive when the valve is open will be reflected as the opposite type (I have not shown these in Figure 2.34 to reduce the confusion).

It is evident that the most important wave that arrives just as the valve is closing is the one that has been reflected from the open end of the air intake, as we designed. If by chance one of the multiply-reflected waves arrives at the same time, it will probably have been reflected so often, and will therefore have lost so much strength as a result, that its contribution will be small. We thus have a term similar to Equation 2.41, but with the length being the distance from the valve to the open end of the air intake.

There is always the possibility that several multiply-reflected waves of the same type will arrive at approximately the same time, and will therefore add and produce a large effect. Such a thing has happened in Figure 2.34, in the compression wave of strength 0.922, which arrives about 36 CAD before the valve closes. This is a constructive interference between two waves, both resulting from the opening of the number 1 inlet valve, but having reflected in a different pattern. In this case, this arrival will have no effect, since the valve is still open. However, the engine could have been designed to take advantage of this, changing the lengths slightly so that the valve was just closing when this wave arrived.

Finally, I show in Figure 2.35 the situation in a 120 degree V6 engine, in which there is substantial overlap of the open period of the valves in different cylinders, since a cylinder fires every 120 CAD. This complicates the situation still further, but the principle remains the same – only one wave can arrive just as the inlet valve is closing, and you will have to choose which one you want.

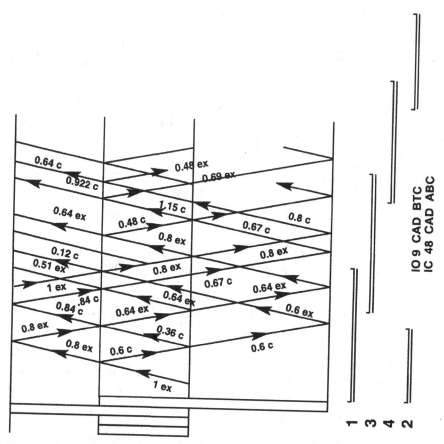

Figure 2.34. $x - t$ diagram of waves in the intake manifold of a four-cylinder engine, with a longer air intake than shown in Figure 2.32, and tuned for the return of the wave from the end of the air intake. The nature and intensity of the waves is indicated on each ray.

2.12.4 What Does ESP Do?

The Helmholtz resonator approach and the organ pipe approach are simple physical models that help us to understand the basic physics of manifold flow. However, when it comes to calculating numbers that might be useful in designing a tuned manifold, they fall considerably short of our needs. There are too many crude approximations, which make the numbers quite inaccurate. ESP does something considerably more sophisticated than the Helmholtz resonator approach and the organ pipe approach, which embodies more-or-less the same basic physics without the crude approximations, and is therefore far more accurate. Wave propagation *and* inertial sloshing in the manifold are handled almost exactly by the method of characteristics. This is a non-linear approximation, in which conserved quantities are followed as they propagate on wave fronts. In this way the accumulation of mass in the manifold due to compressibility of the gas is dealt with, something that neither the organ pipe approach nor the Helmholtz resonator approach can manage. ESP also treats almost exactly the propagation of waves (and the accumulation of mass) in the inactive manifold branches, using

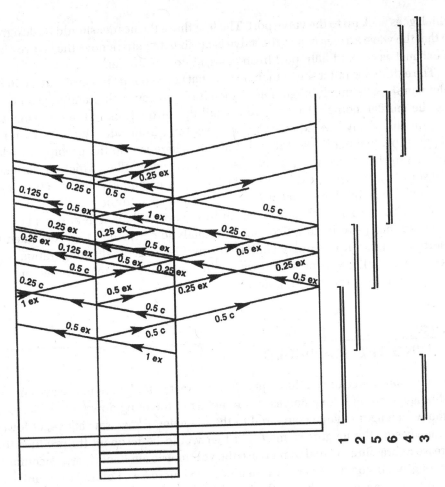

Figure 2.35. $x - t$ diagram of a V6 engine, firing at 120 CAD.

the physics of the Helmholtz resonator approach without using it directly, and of course it treats exactly the isentropic flow through the valve and the accumulation of mass in the cylinder. ESP does include plenums corresponding to the air filter and muffler/catalytic converter, and compressible flow into and out of these is handled exactly, again without having to use the simplifying Helmholtz resonator approach. All this is explained fully in Chapter 8.

2.12.5 The Exhaust System

The situation in the exhaust system is similar to that in the inlet system. The temperature of the exhaust is about 1300 K. Since the speed of sound is $a = \sqrt{\gamma RT}$, the speed of sound is higher by a factor in the neighborhood of 2. This affects natural frequencies and lengths.

In the exhaust system, it is a compression wave that is emitted when the valve first opens. It runs down to the branch point of the manifold, where it is reflected as an expansion wave of strength 0.6 (if we are considering a four-cylinder engine), which goes back up to the valve port. The lengths of the header should be designed so that the wave arrives just as the valve is closing, to help remove the last vestiges of exhaust gases, and help pull fresh charge into the clinder.

The exiting leg (after the branch point), that corresponds to the air intake in the inlet manifold, is much longer here, since it runs through the catalytic converter and the muffler, going ultimately to the tailpipe at the back of the car. This will considerably delay the expansion wave resulting from reflection at the end of the tail pipe, making it absolutely irrelevant. There is also the problem of what happens to the waves when they pass through the muffler, which is designed to absorb such waves. I expect that there will be essentially no reflection from the muffler termination – that waves will disappear into the muffler and never emerge. This is like an absorbing beach in a wave tank, which absorbs surface waves, so that there is no reflection. The catalytic converter is not designed to absorb sound, but it is crudely similar to a muffler in general configuration, and will act in a similar way on sound waves. In any event, we are interested only in the reflection from the branch point.

2.13
FOLDING THE MANIFOLD

The inlet manifold can be folded, just as a trumpet is folded. Sound waves in the mid-range of audible frequencies ignore bends in traveling down a tube, as long as the wavelength is large compared to the diameter. As we saw above, at lower engine speeds, the lengths of manifold that would be resonant (to produce the increase of pressure needed to increase the volumetric efficiency), are uncomfortably long, but become shorter as the engine speed increases. Folding the manifold, and including a valve to change the length, would solve this problem, providing a relatively long manifold at low speeds, and a shorter manifold at higher speeds. In Figure 2.36 I have sketched such a manifold.

This folded section could be the air intake, so that it would serve for all cylinders.

In Figure 2.37 I show a cutaway of one cylinder of the new Mercedes-Benz M-Class V6 engine. Note that this engine has two spark plugs, which shortens the combustion time, reducing time losses for an increased η_i (see Chapter 1), as well as permitting a somewhat higher compression ratio without knock (since it reduces the time for the autoignition reaction to take place). The plugs are not fired simultaneously, which reduces the initial slope of the pressure-rise curve, to reduce combustion noise. It has three valves per cylinder: two intake and one exhaust, which is not an absolutely optimum arrangement (See Section 2.8), but is there to produce tumble (see Chapter 5). It has roller followers, which reduces frictional losses to the valve train (see Section 2.9). And it has a very long folded intake system with a flap valve to shorten the path for high speed operation. As a result, it achieves a specific output of 0.42 kW/cm^2(= 3.64 HP/in^2), and 50 kW/L(= 67.2 HP/L) at a piston speed of 15.4 m/s(= 3080 ft/min), and a *bmep* = 1.09 MPa(= 159 psi) (compare the figures given in Chapter 1).

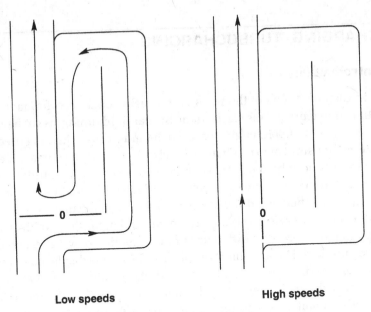

Low speeds   **High speeds**

Figure 2.36. Sketch of a folded manifold with valve.

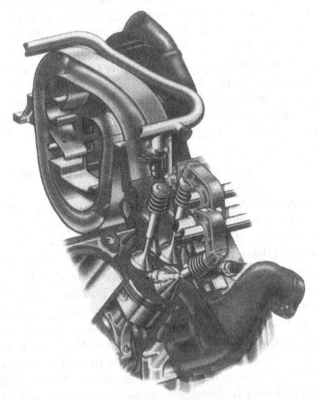

Figure 2.37. Mercedes-Benz M-Class engine. Note two spark plugs, three valves per cylinder, roller followers, long folded intake runner with flapper valve [28]. © Mercedes Benz of North America. Inc.

2.14

SUPERCHARGING/TURBOCHARGING

2.14.1 Introduction

As we saw in Chapter 1, one of the ways of increasing engine performance is by increasing the inlet density. This can be done by manifold tuning, as we have seen in Section 2.12.3, or by supercharging or turbocharging. Supercharging used to be the generic term for using some mechanical device for increasing the inlet density; now it is taken to refer only to positive displacement devices, and turbocharging is used to refer to dynamic devices. A dynamic device is one, like a centrifugal compressor, in which there is a direct connection between the inlet and the outlet (so that, if the device were not rotating, there would be no pressure difference), and which produces its pressure difference from the motion of winglets, or blades, through the air, produced by rotation of the device. A positive displacement device has no direct connection between inlet and outlet; air is taken into a chamber, which is then closed. The volume of the chamber may be reduced, increasing the pressure, and the chamber is then opened to the outlet. In Figure 2.38 we show sketches of several of the most popular superchargers and turbochargers. The Paxton-McCulloch and Latham are dynamic compressors, while the Judson and Roots are positive displacement machines. The Paxton-McCulloch is centrifugal, while the Latham is axial flow, like the compressor in a jet engine. The rotors in the Roots are not in contact, but are held in the proper relative position, and are rotated, by gears outside the compressor casing. Notice that the air trapped between the rotors in the Roots is not compressed (since the volume of the space does not change), until it is delivered to the receiver at the outlet, when it is compressed by the backflow of the higher pressure air in the receiver. For this reason, the efficiency of the Roots is a little lower than that of other types. The vanes in the Judson, on the other hand, slide in and out, changing the volume of the spaces as they carry the air around from inlet to outlet.

Historically, probably the best known example of the use of a Roots type blower is the so-called Blower Bentley (see Figure 2.39). This lovely car has a Roots blower mounted on the front end of the crankshaft, which can be seen in between the frame horns, in front of the radiator in the Figure. It is fed by a pair of side-draft SU carburettors which can also be seen. A car like this sold in 1996 for a figure in the neighborhood of half-a-million dollars. In fact, W. O. Bentley is reputed not to have approved of the use of the blower on his car, although it had great market appeal.

Another notable use of the Roots-type blower was on post-war General Motors diesel buses, which were two stroke and scavenged by General Motors manufactured Roots-type blowers. These engines were made in various numbers of cylinders, three, four, five and more, depending on the service. The blowers tended to outlast the motors, and were recycled, appearing notably in racing cars of various sorts. In Figure 2.40 I show a fuel dragster, designed to operate on exotic fuel, to produce the maximum speed after a standing quarter mile. The large device on top of the flat head V8 is the GM Roots-type blower.

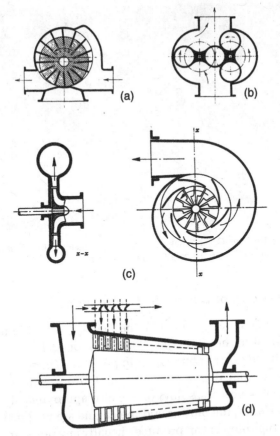

Figure 2.38. Several popular superchargers and turbochargers. (a) is a Judson vane type from about 1950,
used on the MGA and VW. (b) is a Roots type, used
by GMC and recycled (see text) on Fuel Dragsters.
(c) is a Paxton-McCulloch, typical of modern centrifugal turbochargers. (d) is a Latham, used on '50s Ford
V8s. Copyright © 1968 and new material © 1985 by
The Massachusetts Institute of Technology.

Figure 2.39. A Blower Bentley. Reproduced by kind
permission of Stanley Mann, The Fruit Farm, Hertford-
shire, England.

Figure 2.40. The Greer-Black-Prudhomme dragster [36]. Photo: Steve Coonan © courtesy *The Rodder's Journal.*

2.14.2 Characteristics of Super/Turbochargers

A positive displacement supercharger, like the Roots, produces a mass flow rate essentially proportional to the shaft speed, because for each turn of the shaft it ingests the same volume of air at atmospheric pressure. There is a little leakage at the higher pressures, but it is negligible. The blower shaft is geared to the crankshaft, since the engine's mass flow rate is also approximately proportional to the engine speed. Hence, the blower and the engine are well matched in mass flow rate. The Roots-type blower can produce virtually any pressure rise at a given speed. Ignoring for a moment the volumetric efficiency, which will increase with the use of a blower, the blower mass flow must equal the engine mass flow. If N_e is the engine speed, V_d the engine displacement, ρ_i the inlet density (after the blower), V_b the blower displacement, N_b the blower speed and ρ_a the atmospheric density, then we have

$$V_b N_b \rho_a = V_d \frac{N_e}{2} \rho_i \tag{2.43}$$

which can be rearranged to give

$$V_b = V_d \frac{N_e}{2 N_b} \frac{\rho_i}{\rho_a} \tag{2.44}$$

and the pressure can be calculated from the density and temperature. The blower is geared to run faster than the engine, but proportional to it.

The behavior of a turbocharger is quite different. The pressure difference between the top and bottom of a wing is proportional to the square of the speed (due to Bernoulli's equation [31]); because a dynamic compressor consists of an array of blades, which are just small wings, the pressure difference generated by a turbocharger is just the pressure difference across the array of blades, and hence is also proportional to the square of the rotational speed. The mass flow rate is not tied to the rotational speed by a fixed displacement; at a fixed rotational speed, as the pressure difference increases, the mass flow rate decreases. I have included here plots (Figures 2.41, 2.42) of the characteristics of positive displacement and dynamic compressors, for comparison.

Note how the positive displacement blowers have nearly vertical lines of constant tip speed, producing nearly any pressure ratio at the same mass flow rate and the same tip speed. The dynamic turbochargers, on the other hand, require a substantial increase in mass flow for an increase in pressure ratio at approximately constant efficiency. Note also the line labeled surge line. The dynamic machines display a fluid mechanical instability under certain operating conditions; as a result, they cannot be operated to the left of the surge line.

This difference in basic behavior results in very different performance of the two types of blowers. In the case of the positive displacement blowers, the density ratio is essentially fixed by the ratio of the displacements (and rotational speeds) of the engine and the blower. As a result, the *bmep* will be increased about the same amount at all engine speeds. With a dynamic machine, on the other hand, the *bmep* obtained will rise rapidly with engine speed. In Figure 2.43 we show a typical curve of *bmep* for a naturally aspirated and turbocharged engine. Note how the curve of *bmep* for the turbocharged engine rises strongly until perhaps 2,200 rpm. This is the rising characteristic due to the dependence of the pressure

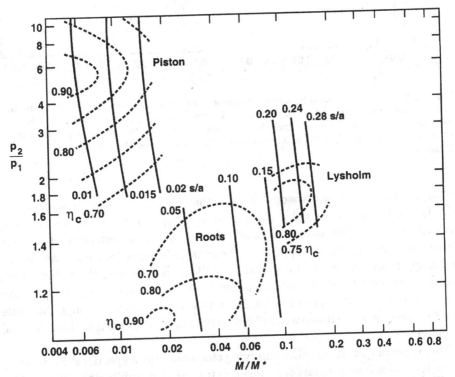

Figure 2.41. Characteristics of positive displacement compressors. From [94]. *s* is the tip speed of the rotor, and *a* is the speed of sound. η_c is the thermodynamic efficiency of the compressor. $\dot{M}^*$ is the mass flow at choked (sonic) conditions through an orifice with a diameter equal to that of the rotor or piston. Copyright © 1968 and new material © 1985 by The Massachusetts Institute of Technology.

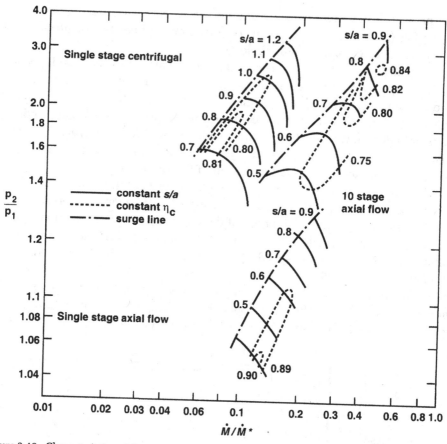

Figure 2.42. Characteristics of dynamic compressors. From [94]. s is the tip speed of the rotor, and a is the speed of sound. η_c is the thermodynamic efficiency of the compressor. $\dot{M}^*$ is the mass flow at choked (sonic) conditions through an orifice with a diameter equal to that of the rotor. Copyright © 1968 and new material © 1985 by The Massachusetts Institute of Technology.

on the square of the shaft speed. The fact that the *bmep* curve flattens at that point is due to what is called a waste gate or dump valve. The turbocharger is deliberately made with excess capacity, so that it will be effective at a lower engine speed. As the engine speed rises, the boost (additional) pressure, and hence the *bmep*, produced by the turbocharger will quickly exceed the value which produces knocking. At the maximum allowable value, further rise in the pressure is prevented by venting the additional mass flow to atmosphere through the waste gate. At higher rpm, the *bmep* will fall due to the fall in volumetric and mechanical efficiencies.

The turbocharger has another annoying characteristic. At partial throttle when power levels are low the mass flow through the exhaust turbine is relatively low, and the turbine and compressor (since they are on the same shaft) speed is relatively low, and hence the boost is small. If the throttle is suddenly opened, the turbine/compressor shaft speed does not respond instantaneously, and it can take times of the order of a second or two to produce boost. This can seem like a very long time in certain situations. This is known as turbo lag.

2.14.3 Thermodynamic Considerations

We can apply the first law of thermodynamics, in the form of a steady flow energy equation, to a control volume around a blower or compressor or turbine:

$$\dot{Q} - \dot{W} = \dot{m}\left[\left(h + \frac{v^2}{2} + gz\right)_{out} - \left(h + \frac{v^2}{2} + gz\right)_{in}\right] \quad (2.45)$$

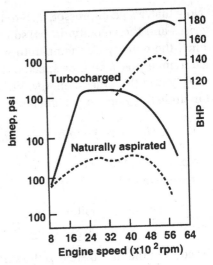

Figure 2.43. Full-throttle performance of 1982 spark-ignition automobile engine, turbocharged and normally aspirated. Engine is vertical 4-cylinder, 86.1 mm bore, 78.7 mm stroke, 2.75 L, with compression ratios of 7.4 (turbocharged) and 8.8 (normally aspirated) [69]. Reprinted with permission from 820442 © 1982 Society of Automotive Engineers, Inc.

where $\dot{Q}$ is the heat transfer rate into the control volume, $\dot{W}$ is the shaft work transfer rate out of the control volume, $\dot{m}$ is the mass flow rate, h is the enthalpy per unit mass, $\frac{v^2}{2}$ is the kinetic energy per unit mass and gz is the potential energy per unit mass.

Many of these terms are negligible and can be omitted. $\dot{Q}$ for pumps, blowers, compressors and turbines is usually small enough to be neglected [47]. The difference in potential energy gz is not important.

The kinetic energy per unit mass is also usually negligible. This can be seen in the following way: the isentropic speed of sound is $a = \sqrt{\gamma RT}$, so that

$$\Delta h = c_p \Delta T = c_p \frac{\Delta a^2}{\gamma R} = \frac{\Delta a^2}{\gamma - 1} \quad (2.46)$$

As a result, the ratio of the change in kinetic energy to the change in enthalpy becomes

$$\frac{\Delta(v^2/2)}{\Delta(h)} = \frac{\gamma - 1}{2} \frac{|M_o^2 a_o^2 - M_i^2 a_i^2|}{|a_o^2 - a_i^2|} \approx O(M^2) \quad (2.47)$$

where M_o is the outgoing Mach number and M_i the ingoing Mach number, and a_o the outgoing speed of sound and a_i the ingoing speed of sound, of the flow in the compressor or turbine. M is either incoming or outgoing, the point being that there is little difference between them. Under the most extreme circumstances, the Mach number is probably not above 0.3, and is usually much lower. In a typical situation, with $M \approx 0.3$, the right-hand side of 2.47 is of order 6%, and becomes small in proportion to M^2 as the Mach number is reduced.

Note that, in a compressor, $a_o > a_i$ because the temperature rises, and the density rises and the velocity drops, so that the Mach number drops, $M_i > M_o$. In a turbine, the density and temperature go down and the velocity up, so that the Mach number rises, and the speed of sound drops.

Neglecting the kinetic energy, the thermodynamic shaft work out (this does *not* include the mechanical efficiency; we will deal with that in a moment) is

$$-\dot{W} = \dot{m}(h_{out} - h_{in}) = \dot{m}c_p(T_{out} - T_{in}) \tag{2.48}$$

(since c_p is essentially constant for air, or fuel-air mixture at inlet temperatures). Finally, we define a compressor isentropic efficiency as

$$\eta_C = \frac{\text{adiabatic reversible power requirement}}{\text{actual power requirement}} \tag{2.49}$$

(applied from the initial state to the *same final pressure*). We can thus write η_C as

$$\eta_C = \frac{h_{2s} - h_1}{h_2 - h_1} \tag{2.50}$$

where h_{2s} is the enthalpy at the exit pressure if the process had been isentropic, while h_2 is the actual enthalpy at the exit pressure. h_1 is the enthalpy at the entrance state.

Equation 2.50 becomes

$$\eta_C = \frac{T_{2s} - T_1}{T_2 - T_1} \tag{2.51}$$

In an isentropic process, the inlet and outlet temperatures and pressures are related by

$$\frac{T_{2s}}{T_1} = \left(\frac{p_2}{p_1}\right)^{(\gamma-1)/\gamma} \tag{2.52}$$

Combining these, the power required to drive the compressor (from a thermodynamic point of view) is

$$-\dot{W}_C = \frac{\dot{m}c_p T_1}{\eta_C}\left[\left(\frac{p_2}{p_1}\right)^{(\gamma-1)/\gamma} - 1\right] \tag{2.53}$$

Equations 2.48 or 2.53 are just the work required thermodynamically. There are also mechanical losses in the blower or compressor. These are primarily bearing friction losses. The power required to drive the device, say $-\dot{W}_{C,D}$, is given by

$$-\dot{W}_{C,D} = -\frac{\dot{W}_C}{\eta_m} \tag{2.54}$$

where η_m is the mechanical efficiency of the compressor or blower. The two efficiencies appear as a product $\eta_C\eta_m$, and are difficult to separate in practice.

They are usually combined into a single efficiency. Small Roots-type blowers have peak overall efficiencies in the neighborhood of 0.55 (see [47]). Single stage centrifugal compressors have peak efficiencies in the neighborhood of 0.7 (see [47]).

2.14.4 Turbines

For the exhaust gas turbine, the isentropic efficiency is defined as

$$\eta_T = \frac{\text{actual power output}}{\text{adiabatic reversible power output}} \qquad (2.55)$$

(applied from the initial state to the *same final pressure*). That is, the inverse of the definition for the compressor, because the turbine produces power, instead of absorbing it. In exactly the same way as for the compressor, we can write

$$
\begin{aligned}
\dot{W}_T &= \dot{m}(h_3 - h_4) \\
&= \dot{m}c_p(T_3 - T_4) \\
&= \dot{m}c_p\eta_T(T_3 - T_{4s}) \\
&= \dot{m}c_p\eta_T T_3\left(1 - \frac{T_{4s}}{T_3}\right) \\
&= \dot{m}c_p\eta_T T_3\left[1 - \left(\frac{p_4}{p_3}\right)^{(\gamma-1)/\gamma}\right] \qquad (2.56)
\end{aligned}
$$

where 3 refers to the state at the inlet to the turbine, and 4 to the state at the exit. Enthalpy is dropping, since work is being extracted. All the constants (c_p and γ in particular) must be evaluated at the exhaust gas temperatures. Since the speed of sound is nearly doubled at exhaust gas temperatures, the Mach number will be even lower, and the neglect of the kinetic energy is even more justified.

Again, there is a mechanical efficiency, and the actual work output of the turbine is

$$\dot{W}_{T,D} = \eta_m \dot{W}_T \qquad (2.57)$$

where η_m is the mechanical efficiency. In the same way, it is usually impossible to separate the mechanical from the thermodynamic efficiency, and a single figure, the product of the two, is usually given. Single stage radial flow turbines have peak overall efficiencies in the neighborhood of 0.7 (see [47]).

2.14.5 Knock

The use of supercharging is complicated by the problem of knocking. As the inlet density increases, peak cylinder pressures increase. A measure of this is the increase in the *bmep* and *imep*. Knocking is discussed in detail in Chapter 1. Figure 2.44 ([10], [94], reproduced from [94]) shows knock-limited performance

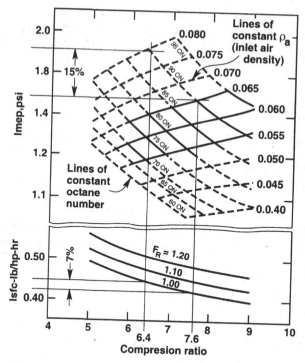

Figure 2.44. Knock-limited *imep* and octane requirement vs. compression ratio and inlet density. ON = fuel octane number; *imep* values are for 1,500 rpm; $F_R = 1.2$; $T_i = 70°$F. Copyright © 1968 and new material © 1985 by The Massachusetts Institute of Technology. Taylor [94] erroneously attributes this figure to Barber [10], where it does not appear. I have not been able to identify a source, so will attribute it to Taylor [94].

of a spark-ignition engine at one set of operating conditions, including best-power spark timing and fuel-air ratio, and constant inlet-air temperature. The values of inlet air density given on the figure are in lb_m/ft^3. At 20°C(= 293 K) and 1.013×10^5 Pa, air density is 0.0749 lb_m/ft^3(= 1.2 kg/m³). At more typical inlet conditions of 0.92 atm = 0.932×10^5 Pa and 322 K, the density is 1.00 kg/m³(= 0.0627 lb_m/ft^3). With turbocharging, if $p_2/p_1 = 1.4$ (using Equation 2.53), starting from the same temperature, the density is 1.14 kg/m³(= 0.0856 lb_m/ft^3). Thus, the figure barely covers densities corresponding to mild turbocharging without intercooling, the lower part of the figure corresponding to partial throttle operation. However, the principle is not compromised.

The curves labeled, for example, "90 ON" are for constant octane number. The following definition of octane number is loosely paraphrased from [99]: "The octane number defines a gasoline's antiknock quality, its resistance to preignition. The higher the octane rating, the greater the resistance to engine knock. Two different procedures are in international use to determine the octane rating: the Research Method (producing the Research Octane Number, RON) and the Motor Method (producing the Motor Octane Number, MON).

In both methods, a test engine is run with the fuel in question, and its knock resistance is compared with that of a mixture of iso-octane (C_8H_{18}) and n-heptane (C_7H_{16}). Iso-octane is extremely knock resistant, and is assigned an octane number (RON or MON) of 100; n-heptane has a very low knock resistance, and is assigned the number zero. The octane rating (by either method) is the percentage of iso-octane that produces the same knock resistance as the fuel in question in the test engine.

The Motor Method differs from the Research method by using preheated mixtures, higher engine speeds and variable ignition timing, placing more stringent thermal demands on the fuel in question. MON figures are lower than RON figures."

At the pump, the average of the two is usually given, designated by $(R+M)/2$. In Figure 2.44 the ON given is the $(R+M)/2$.

Suppose that your engine is operating with 90 ON fuel at a compression ratio of 7.6, and an inlet density of 0.065 lb_m/ft^3. If you decide to install a turbocharger which raises the inlet air density to a little below 0.80 lb_m/ft^3, you will either have to use a fuel with about 97 ON, or retard the spark substantially to limit the peak pressure, or reduce the compression ratio to 6.4 to allow you to continue to use the same fuel and timing. If you decide to reduce the compression ratio (which will require a different cylinder head), you will get approximately a 15% increase in *imep* at the cost of a 7% increase in specific fuel consumption. Engines that are designed for turbocharging are designed with lower compression ratios for this reason. Also, engine control computers for cars designed with turbocharging retard the spark in response to input from a knock sensor, which senses structural vibration of the engine block in response to the rapid pressure fluctuations associated with knocking (see Chapter 1).

2.15
INTERCOOLERS

Suppose your Blower Bentley has a Roots-type blower with an overall efficiency of 0.55, operating at a $M \ll 1$. The inlet temperature is 293 K, and the inlet pressure is 1×10^5 Pa. The blower produces a boost of about half an atmosphere, giving an outlet pressure of 1.5×10^5 Pa. Using Equation 2.53, we find that the temperature rise is $T_2 - T_1 = 65.4°C$, giving a density of $\rho = 1.45$ kg/m^3.

This density could be considerably increased if the temperature of the air could be reduced before it enters the engine. This is the function of the intercooler.

The intercooler is a heat exchanger, similar to the radiator, that cools the air after it has passed through the compressor or blower. The heat is either exchanged directly with the atmosphere, or is exchanged with a liquid coolant, which is itself cooled in another radiator, exchanging heat with the atmosphere. How this is accomplished is dictated by convenience of location. In Figure 2.45 I show a Renault V6 Formula 1 engine from 1978, turbocharged and intercooled, DOHC, four valves per cylinder. This had a bore of 86.2 mm and a stroke of 42.8 mm,

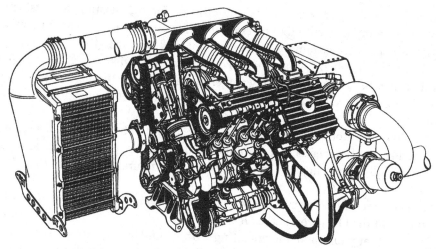

Figure 2.45. Renault 1.5 L turbocharged V6 for Formula 1 from 1978, developing 380 kW (510 HP) at 10,800 rpm. Note the large intercooler (left) [17]. Reproduced by kind permission of the estate of Griffith Borgeson.

surely something of a record stroke:bore ratio at $S/b = 0.497$. The engine achieved a specific output of 1.09 kW/cm^2 (= 9.36 HP/in^2), and 253 kW/L(= 340 HP/L), at a piston speed of 15.4 m/s. The turbocharger is the small device at the right-hand side, to which the exhaust pipes lead, while the intercooler is the large device on the left. This is an air-to-air intercooler.

Another intercooler is shown in Figure 2.46. This is a liquid-to-air radiator, which is placed underneath the rear deck, with an air intake under the downforce wing.

Finally, in Figure 2.47 I show the Oshkosh Phoenix off-road fire control vehicle. This weighs 64,500 lbs. fully loaded, and can climb a 60% grade. It is all-wheel drive, and the two front axles steer. In Figure 2.47, the man is pointing to a window in the floor of the vehicle, which the driver must use when the vehicle is climbing such a grade. I have included this vehicle because it is a supercharged *and* turbocharged and intercooled SOHC 16-valve 2-stroke diesel V8 of 12.054 L displacement. This develops 332 kW(= 445 HP) at 2,100 rpm, for a specific power of 27.5 kW/L(= 36.9 HP/L). It produces a *bmep* = 0.787MPa(= 114 psi). *Car and Driver* does not give information on bore and stroke, but assuming that $b = S$, the mean piston speed is 4.34 m/s(= 868 ft.min), and the power/piston area is 0.344 kW/cm^2(= 2.96 HP/in^2). Such an engine, designed for extreme reliability and long service, is very much under-loaded compared to a Formula 1 engine, particularly with regard to piston speed and *bmep*. However, the power per unit piston area is quite respectable. Many truck engines are turbocharged and intercooled, since this produces the same power from an engine which weighs substantially less. In this case, because it is a two-stroke diesel, the super- and turbocharging also serves a scavenging function. The supercharger in particular provides scavenging at low speeds when the turbocharger is providing little boost.

The performance of an intercooler is specified by a figure of merit called the effectiveness, C_c:

$$C_c = \frac{T_1 - T_i}{T_1 - T_w} \qquad (2.58)$$

where T_1 is the temperature of the charge leaving the compressor or blower and entering the intercooler, and T_i is the temperature of the charge leaving the intercooler and entering the engine. T_w is the coolant entrance temperature. Intercoolers are designed with figures of merit between 0.6 and 0.8.

In our example at the beginning of this section, suppose that we use an intercooler with a figure of merit of 0.7, operating in ambient air at 293 K. Using Equation 2.58 we find a new temperature at the engine inlet of 312 K, giving an inlet density to the engine of 1.67 kg/m^3, a considerable improvement.

Figure 2.46. Intercooler of the Porsche 911 Turbo S, underneath the whale tail [13]. Photo by Dick Kelley, courtesy of *Car and Driver*.

Figure 2.47. The Oshkosh Phoenix 8 × 8 fire control vehicle [77]. Photo by Aaron Kiley, courtesy of *Car and Driver*.

2.16
PROBLEMS

1. Working with ESP, using no manifold and any convenient configuration, set the inlet pressure to 0.37 atm (corresponding to idle). Determine peak pressure, and compare with peak pressure when inlet pressure is 0.92 atm. What is volumetric efficiency at 0.37 atm $= p_i$, and at 0.92 atm $= p_i$?

2. • Estimate by graphical or other means the relative sizes of five equal sized valves (three intake, two exhaust) fitting in a given bore circle of diameter b compared to two valves (exhaust 3/4 diameter of the intake) in the same cylinder. Assume that there is no spacing among the valves and the cylinder bore.

   • Then find the increase in total A_{flow} for lift $D/4$ and what fraction higher the piston speed can be to have the same volumetric efficiency. Note that the equation for Z given in the text assumes one intake valve. You will have to modify it for three intake valves, accounting for the increased inlet area.

   • If the engine can operate at the higher piston speed with the same mechanical and other efficiencies, what would be the increase in power?

3. A stock inlet cam profile is open for 240 CAD. The cam profile is ground so that the valvetrain has a uniform upward acceleration (say, a) for 40 CAD, a uniform downward acceleration of half this value ($a/2$) for 160 CAD, and closes on a uniform upward acceleration (again a) for 40 CAD. This is a pushrod actuated, OHV engine. The valvetrain has a mass of 0.50 kg. The valve spring, compressed to operating length, exerts a force of 705 N. The valve lift $L = 8.5$ mm. The relation between valve lift and acceleration for this cam profile is $L = aT^2/24$. At what engine speed will valve float begin?

4. Work with ESP, using lumcfrx.ess as setup file (this gives a CFR engine with a 4.92 compression ratio). The valve settings are: IO $= 357$ CAD ACTC (After Compression TC), IC $= 573$ CAD ACTC, EO $= 120$ CAD ACTC, EC $= 363$ CAD ACTC. The profile is a cosine form. These settings do not allow for the ramp, which is approximately the first and last 10% of the cosine curve. That is, the settings should really be: IO $= 331$ CAD ACTC, IC $= 599$ CAD ACTC, EO $= 91$ CAD ACTC, EC $= 392$ CAD ACTC. Change the valve settings to these new figures. What effect does this have on the volumetric efficiency? On the indicated efficiency?

5. Given a 3.0 liter six-cylinder engine with $D_{in} = b/2$ and $D_{ex} = 0.87\,D_{in}$, compression ratio $r = 9.5$, air cycle thermodynamic efficiency $\eta_{ac} = 1 - (1/r)^{k-1}$ ($k =$ specific heats ratio $= 1.4$), real gas thermodynamic efficiency $\eta_o = 0.65\,\eta_{ac}$, indicated thermodynamic efficiency $\eta_i = 0.86\,\eta_o$, mechanical efficiency $\eta_m = 0.85$, average $C_i = 0.35$, bore/stroke $= 1.06$, fuel air ratio $= 1/14.6$,

$Q = 40.6$ MJ/kg, air density $= 1.2$ kg/m^3. Assume the speed of sound $a = 343$ m/s.

- At 5,500 rpm, find the volumetric efficiency and the output power using straight line approximations: volumetric efficiency $= \eta_v = 0.83$ for $Z < 0.5$; $\eta_v = 1.03 - 0.4Z$ for $Z > 0.5$.
- Find the volumetric efficiency and the power output at 2,000 rpm.

6. Given the following data for a Jaguar XK engine: displacement $V_d = 3.4$ L, 6 cylinders, 157 kW @ 5,500 rpm, 343 Nm of torque @ 2,310 rpm, intake port diameter $D = 38.1$ mm, exhaust port diameter $D = 34.3$ mm, bore $b = 83$ mm. Valve lift $L = 9.5$ mm. Use $k = 1.4$ (ratio of intake open angle to 180 CAD). Assume $a = 343$ m/s.

- Figure 2.33 is actual test data for a similar (but higher performance) engine. Work with ESP, using XK348.ess as setup file. This gives an XK engine with 3.4 L displacement and $r = 8$. For each of the intake lengths on Figure 2.33, use ESP to determine the expected peak ramming rpm. Assume each cylinder has a separate intake stack with WOT, and no EGR. Assume the stack is the same diameter as the intake port.
- Assume that the exhaust consists of equal length headers, length $L = 1.7$ m, combined into a single tail pipe. Suppose that the headers are the exhaust port diameter, and the tail pipe has the combined area of the headers. Assume no muffler or catalytic converter.
- Compare the results from the first two parts to Figure 2.33 and comment on the level of agreement.

7. • The engine in problem 5 is turbocharged so that at full throttle, the air density out of the compressor is 1.7 times atmospheric. Assuming ideal isentropic compression, find the output pressure of the compressor and the resulting speed of sound. Then find the power with the new air density and Mach index at 5,500 rpm.

- Repeat the calculations for a compressor with 75% efficiency.
- Repeat the calculations with the compressor in the second part along with an intercooler which cools the compressed air. The intercooler runs at constant pressure, and has an effectiveness of 0.80. Assume coolant temperature is 60 C.

8. Working with ESP, using no manifold and any convenient configuration, set the inlet pressure to 1.1 atm, 1.3 atm, and 1.5 atm (representing supercharging). Determine peak pressures, and compare with peak pressure when inlet pressure is 0.92 atm. What are volumetric and indicated efficiencies at $p_i = 1.1$ atm, 1.3 atm and 1.5 atm?

9. Consider an exhaust header system for a Jaguar XK engine. The cylinders have equal length branches, which come together at the exhaust pipe. When the exhaust valve opens, a compression wave travels down the branch to the

place where the branches come together, and is reflected as an expansion wave. The exhaust valve is open 235°. Exhaust gas temperature is 1,172°K. Using organ pipe theory, how long should the branches be to provide a power boost at 6,000 rpm?

10. Consider an engine of the early 1930s, eight-liter displacement, six cylinders. The first experimental model is found to have its peak power at 3,168 rpm. It has a bore of 94.6 mm and a stroke of 189 mm. Why is the engine probably peaking at this relatively low rpm? As a consultant, what basic change would you suggest to the designer to raise the peak-power rpm? Back up your explanations and recommendations with numbers.

11. You are going to redesign the Mustang 5.0 L V8, which produces 164 kW @ 4,500 rpm. You are going to replace it with a smaller, turbocharged engine which will produce the same power at the same engine speed. Assume that the fuel octane number (85) remains the same. You will have to reduce the compression ratio from 8.1 to 6.8; the turbocharger raises the inlet air density by a factor of 1.22, raising the bmep from 0.88 MPa to 1.0 MPa. What is the new displacement? How much is the overall efficiency reduced? What is the probable cause of the reduction in efficiency?

ENGINE COOLING

3.1
INTRODUCTION

Combustion temperatures are in the neighborhood of 2,500 K in a spark-ignition engine, and the exhaust gas temperature is about 1,300 K. On the other hand, the melting point of aluminum is about 933 K, and the melting point of iron is about 1,808 K. It is clear that some provision has to be made to keep the piston, valves, and cylinder walls cool, or they will melt. Even considerably short of melting, at high temperatures metals begin to lose their strength, and this must also be avoided. Any heat removed represents a loss of energy, so we want to cool only as much as is necessary to maintain the strength of the materials, maintain clearances, and prevent the lubricant from breaking down.

The problem areas are the exhaust valve and the piston crown. The exhaust valve head loses most of its heat to the valve seat (the amount lost to the valve guide is relatively small, because the path is long and the conduction area is small). Unfortunately, when the valve opens, it is exposed to the exhaust gases, which flow past it at high velocities (making for good heat transfer), and while this is happening the valve head is not in contact with its seat.

In water-cooled engines the water returning from the radiator, which therefore has the lowest temperature, usually flows around the cylinders and then up into the cylinder head. This is not the most effective order because the cylinder heads, and particularly the valve seats, require cooling much more than the cylinder walls. However, it is adequate, and cheap.

During the early 1950s at least one engine, the Armstrong-Siddeley Sapphire, had a cooling system in which the return flow from the radiator was directed by a distribution pipe to the head, specifically to the water-jacket side of the exhaust valve seats, so that these would get the best cooling; from there it flowed through the cylinder head, then down to the cylinder walls. Armstrong-Siddeley Motors made primarily aircraft engines, and this probably influenced their design philosophy. More recently, General Motors' Powertrain Division announced the totally redesigned new generation 5.7 L Small Block V8 engine, developed for use in the Chevrolet Corvette [62]. This has a much-touted "reverse flow cooling," which is exactly the same as that in the Armstrong-Siddeley. The author of [62] justifies this type of cooling system by pointing out that a cooler head permits greater spark advance (in cooler gas, the auto-ignition reaction is slower to go to completion,

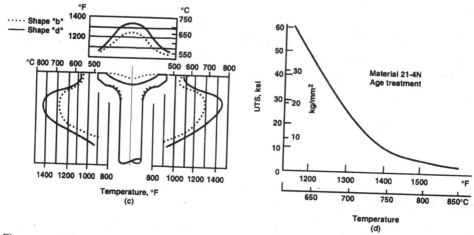

Figure 3.1. Effect of valve-head design on exhaust-valve temperatures, and effect of temperature on strength of an exhaust-valve steel. From [26]. Reprinted with permission from 650484 © 1965, Society of Automotive Engineers, Inc.

so that the end gas is burned before it autoignites), and warmer cylinder bores (within limits) give lower ring friction, both of which are certainly true.

In Figure 3.1 I show temperature distributions for two different shapes of exhaust valves, as well as a plot of ultimate tensile strength versus temperature for an exhaust valve material (from [95]). The temperatures shown in Figure 3.1 are typical. Note that at say, 1,000 K, the valve material has lost a substantial fraction of its strength.

The piston crown is exposed to the gases at combustion temperatures, as is the wall of the combustion chamber. However, the combustion chamber wall has the cooling water on the other side (or the cooling fins in the case of an air cooled engine). The piston crown usually does not have any cooling on the under surface (although it may have oil or water jets, which we will discuss in a moment). Any heat absorbed by the crown must be conducted to the sides of the piston in contact with the cylinder walls, and this path is relatively long. This contact with the cylinder wall is not good from a heat transfer point of view, since through most of the cycle there is no metal-to-metal contact (the surfaces are separated by a thin oil film) and the contact area is small, but the path length is the major factor.

Air-cooled engines run at a higher temperature than liquid-cooled engines (which are limited by the boiling point of the coolant, even under pressure). This makes air-cooled engines somewhat more efficient, but also places them closer to dangerous temperatures. The Volkswagen air-cooled four-cylinder horizontally opposed engine, used in the beetle and the Transporter (or minibus), was one of the most successful large-volume commercial engines of all time. Owners were cautioned to pay attention to valve adjustment. If the valve clearance were allowed to close up, so that the exhaust valve did not completely seat, then it could not transfer heat to the valve seat; in addition, hot exhaust gases would rush through the gap, and for both these reasons, the valve head temperature would rapidly rise.

The oil cooler in the fan housing somewhat blocked the air flow to the forward cylinder on the left, and consequently temperatures were always slightly higher in this cylinder. If the valves had not been adjusted according to the maintainance schedule, and the exhaust valve in this cylinder was being held open because it had no clearance, it would fail at highway speeds (when the heat load was greatest), the valve head would separate from the stem at the neck, fall into the cylinder, shatter the piston crown, and that would be the end of the engine. This was the most common failure mode for this engine, particularly in the United States where frequent routine valve adjustments run counter to our national character.

I will show you shortly that the heat transfer coefficient to the piston crown (and to everything else) is proportional to the piston speed. This means that, as the engine goes faster, temperatures rise. The VW engine was fairly safe from this point of view. However, the Transporter was quite underpowered, particularly when fully loaded with a family on vacation. It was a great temptation to go downhill with wide-open throttle so as to gain enough momentum to get up the next hill. This had the effect of running the engine at wide-open throttle above the maximum design speed. If this was continued for a while, the piston crowns reached their melting point (usually the front left going first), and sprayed molten aluminum all over the inside of the engine. I lost two engines that way.

In many engines the connecting rod is made in such a way, that lubrication oil under pressure is allowed to escape through a small diameter hole, forming a jet which is directed at the underside of the piston crown. This serves to remove heat from the crown. Large diesels, such as might be used on an oil tanker, having bores in the neighborhood of a meter, often use water jets for the same purpose. The larger the piston, of course, the farther the heat must travel to escape from the crown, and the more serious is the cooling problem. Such engines have windage trays which separate the upper crankcase from the oil pan (and on which one can walk into the crankcase when the engine is not running), and it is relatively easy to separate the water from the oil on these trays.

3.2
VALVE SEAT RECESSION

This is a good place to discuss this phenomenon, because the mechanisms involved are related to our discussion in Section 3.1.

In the past Tetraethyl lead was added to all gasoline, primarily to increase the octane rating. However, it had the secondary effect of preventing the valve heads from welding themselves to the seats. If an engine which has never used gasoline with lead is run on unleaded gasoline, each time that the exhaust valve closes pinpoint welds will be formed between the face and the seat. The next time that the valve opens, at the location of each pinpoint weld a little chunk will be torn out of the weaker material, usually the seat. The valve is designed to rotate a little each time it is lifted and consequently the next time the valve closes, the little chunks are not lined up with the holes from which they came. As a result, they keep the valve from seating properly, and hot gases rush between the face and the seat. This makes the situation worse, because the valve face is now much hotter and the pinpoint welds form more readily. The little chunks oxidize in the high temperatures, but more and more are removed, and the valve seat gradually recedes. Tetraethyl of lead forms a protective coating of lead compounds, so that there is not metal-to-metal contact between the valve face and seat, and welds do not form.

This effect is often referred to as lubrication; the phenomenon does involve separating two metals by a thin film of something else, but the something else (the lead compounds) is not a fluid, and does not flow, so the term is somewhat misleading.

When an engine has been run on leaded gasoline, and then is run on unleaded gasoline, it takes a time of the order of one year of normal use to remove all the lead compounds. This is reponsible for the apparent success in the marketplace of most of the additives, which actually do little or nothing – they have been described as being of as much use as a rabbit's foot in the glove compartment. If the motorist changes from leaded to unleaded gasoline, and tries an additive within the first year, because there is still lead in the engine no valve seat recession will take place, and he will conclude that the additive is working.

The customary cure for valve seat recession is to install stellite valve seats, which are extremely hard. This is routinely done in rural machine shops in areas in which leaded gasoline has been phased out, to give new life to old tractors. New cars are delivered now with hardened valve seats. There is some question whether now the chunks will not be torn out of the valve face, because it is now the softer of the two, and valves of harder material become desirable. It is also clearly wise to keep the valve as cool as possible, since the pin point welds form more readily as the valve head temperature rises.

Even with lead in the gasoline, the maintainance of a good seal between valve and seat depends in large part on keeping the valve temperature as low as possible, to keep oxidation low and to prevent warping of the valve. The last engine designed by W. O. Bentley, the 2.6 L DOHC six cylinder for the 1947 Lagonda, was an exceptional engine for its time. David Brown, the tractor manufacturer and racing enthusiast, bought Aston Martin because he wanted to build racing cars and bought Lagonda because he wanted Bentley's engine to put in the racing cars. He did this, producing the Aston Martin DB2, which was very successful in racing. In this engine, Bentley made no provision whatever for valve adjustment. To adjust the valves, the ends of the stems have to be ground. So far as I am aware, he never explained why he did this. My explanation is this: he had designed the engine in such a way that the exhaust valves would be kept very cool at all times and hence should not need adjustment in normal use.

There were two factors in the design of the engine which kept exhaust valve temperatures low. It might be thought that the compression ratio was a beneficial factor, at only 6.5. However, we will see in Section 3.4 that this is actually slightly detrimental.

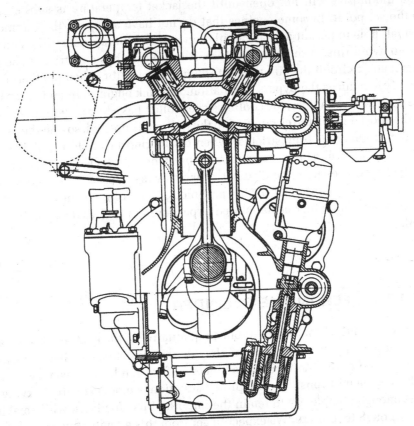

Figure 3.2. The 2.6 L Lagonda six, with barrel crankcase, became the power source of the DB2 Aston Martin in 1950 [17], [42]. Reproduced by kind permission of Aston Martin Lagonda Ltd.

The two factors are: 1) a bypass cooling system and 2) wet valve guides. One speaks of "wet" guides, and "wet" cylinder liners, when the guides, or the liners, are in direct contact with the cooling water. This creates a site for possible leaks between the guide and the block (or the liner and the block), but it greatly shortens the path that the heat must take, even though the valve guide is not usually the major point of heat loss. Wet guides increase the relative importance of the guides in controlling the valve temperatures.

I am including here a cross section of the Lagonda 2.6 L engine, Figure 3.2, showing the wet valve guides.

Engine cooling systems are made with and without a bypass. A bypass permits circulation through the water jacket even when the thermostat is closed. If there is no bypass then the circulation stops when the thermostat closes, and hot spots can develop. These are not normally serious, because temperatures are limited by boiling heat transfer, so long as vapor pockets do not develop. Without circulation, however, the thermostat does not see a representative temperature. This means that the thermostat will not open until the jacket temperature is substantially above the set point. In some engines that do not have a bypass, the thermostat has a small hole to permit a weak circulation past the thermostat even when it is closed, to avoid this. If there is a bypass, it ordinarily has a restriction, so that it will not short-circuit the circulation through the radiator, when the thermostat is open. This reduces the bypass circulation. In truck engines, intended for long service with high reliability, the cooling system often has a bypass which is closed by the thermostat when the circulation to the radiator is opened, so that the bypass circulation is generous (unrestricted) when the thermostat is closed. This keeps the valve seats cool, and helps to avoid warping of the exhaust valve and valve seat wear. The Lagonda engine (see Figure 3.2) had a bypass closed by the thermostat. More recently, the redesigned 5.7 L small block Chevrolet engine intended for the Corvette also has a thermostat-closed bypass system [62], as does the Porsche 928 [48].

3.3
HEAT TRANSFER IN THE CYLINDER

Having discussed the heat transfer problem qualitatively, the next step is to quantify it. What parameters are important? How do they influence the heat transfer?

To understand how the various parameters affect the heat transfer in an engine cylinder, we will consider a simplified situation, a model situation, trying to keep just enough complexity to retain the important mechanisms, while making it simple enough to analyze. We expect to get from this a qualitative answer that will guide us in design.

In Figure 3.3 I have shown a sketch of a valve head, and of a piston crown. You can see that each of them is characterized by a distance L which the heat $\dot{Q}_{cond}$ must travel to the water jacket (we are ignoring the path to the valve guide in the case of the valve, although we could consider that also). In each case the heat passes through material of typical cross-sectional area A_{cond}. The heat from the gases $\dot{Q}_{conv}$ enters through an area A_{conv}.

3.3.1 Conduction in the Solid

The heat transfer problem in the solid (either the valve, or the piston crown) is something like this:

$$\dot{Q}_{cond} = A_{cond}\, \gamma_{solid}\, \frac{\partial T}{\partial x}\bigg|_{solid} \tag{3.1}$$

γ_{solid} is the thermal conductivity of the solid. By way of example, aluminum has a value of $\gamma_{Al} = 204\,\text{W/mK}$ while iron has a value of $\gamma_{Fe} = 81\,\text{W/mK}$. Steel has values in the range $\gamma_{steel} = 48 - 58\,\text{W/mK}$.

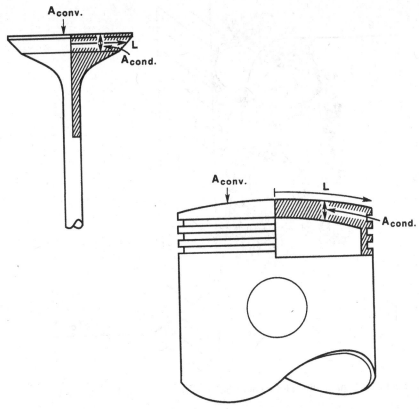

Figure 3.3. Sketch of a valve head, and of a piston crown, showing conduction paths and areas.

$\frac{\partial T}{\partial x}$ is the gradient of the temperature in the solid along the path which the heat takes. We may estimate this in the following way:

$$\frac{\partial T}{\partial x}\bigg|_{solid} = \frac{T_p - T_c}{L} \tag{3.2}$$

That is, we are assuming that the gradient is linear over this distance. T_p is the temperature of the part (valve, piston) with which we are concerned. T_c is the temperature of the sink to which the heat is going (the water jacket, for example). Equation 3.2 allows us to write

$$\dot{Q}_{cond} = A_{cond}\gamma_{solid}\frac{T_p - T_c}{L}. \tag{3.3}$$

3.3.2 Heat Transfer in the Gas

The gas in the cylinder is in turbulent motion. This means that it is in unsteady, irregular, chaotic, swirling, churning motion. This motion transfers heat (and all other properties) several orders of magnitude better than molecular transport (see,

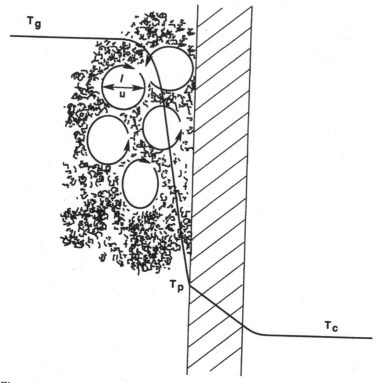

Figure 3.4. Qualitative picture of turbulent heat transfer in the cylinder.

for example, [96]). In Figure 3.4 I have sketched a qualitative picture of the situation. ℓ is the size of a typical eddy, and u a typical fluctuating velocity. We can write in a turbulent fluid

$$\gamma_{conv} = \rho c_p \kappa_{conv} \approx \rho c_p u \ell \tag{3.4}$$

That is, γ_{conv}, the effective convective thermal conductivity, which is due to turbulence, not molecular transport, is proportional to the effective turbulent thermometric diffusivity κ_{conv}, which is roughly equal to the product of the length and velocity scales of the agency which is transporting the heat, the turbulence. When the transporting agency is molecular, the typical length scale is the mean free path, λ, and the typical velocity is the isentropic speed of sound a, and we find that the molecular thermometric diffusivity $\kappa \approx \lambda a$, in exactly the same way.

Now, the convective heat transfer is approximately

$$\dot{Q}_{conv} = A_{conv} \gamma_{conv} \left. \frac{\partial T}{\partial x} \right|_{conv} \tag{3.5}$$

We can estimate the temperature gradient in the gas $\left. \frac{\partial T}{\partial y} \right|_{conv}$ if we look at what causes it. The gas is being mixed by the eddies of size ℓ. Therefore, the profile of temperature cannot have a length scale much different from ℓ, either smaller or larger. So, we expect that

$$\left.\frac{\partial T}{\partial y}\right|_{conv} \approx \frac{T_g - T_p}{\ell} \tag{3.6}$$

where T_g is the mean temperature of the gas a distance from the wall of order ℓ. Because the gas is being well-mixed by the turbulence, we expect that the mean temperature will be fairly uniform in the chamber, except near the wall (say, a few ℓ from the wall). T_p is the temperature of the part in question, either the valve or the piston crown. This means that we can write

$$\dot{Q}_{conv} = A_{conv}\rho c_p u \ell \frac{T_g - T_p}{\ell} \approx A_{conv}\rho c_p u [T_g - T_p] \tag{3.7}$$

This kind of analysis is very approximate. While it tells us how one thing depends on another, and is very useful for comparing design possibilities, we will need to include an adjustable constant to get good numerical values. We usually write

$$\dot{Q}_{conv} = A_{conv} S_t \rho c_p u [T_g - T_p] \tag{3.8}$$

where the constant S_t is the Stanton number, a dimensionless quantity. The combination $S_t \rho c_p u = h_e$, the heat transfer coefficient (e for effective), so that

$$S_t = \frac{h_e}{\rho c_p u} \tag{3.9}$$

3.3.3 Variation of Part Temperature

The situation in the cylinder is never in steady state. The gas temperature rises and falls in each cycle, and the part temperature will also, lagging the gas temperature. However, if we consider cycle averages of all quantities, we can say that

$$\dot{Q}_{cond} = \dot{Q}_{conv} \tag{3.10}$$

We can write

$$A_{conv} S_t \rho c_p u [T_g - T_p] = A_{cond} \gamma_{solid} \frac{T_p - T_c}{L} \tag{3.11}$$

or

$$\frac{T_p - T_c}{T_g - T_p} = \frac{A_{conv} S_t \rho c_p u L}{A_{cond} \gamma_{solid}} \tag{3.12}$$

Equation 3.12 is generally applicable to cylinder walls, piston crowns, or to exhaust valves. It can be used to compare designs, in which sizes and materials may differ. In order to obtain reliable values of T_p, we must use the correct T_g (see Section 3.4) and the correct u (Section 3.3.4). In addition, exhaust valves represent a special case, because they are open for a quarter of the cycle, during which their

A_{conv} is larger and their A_{cond} is smaller, and the gas temperature is higher. I will give this special consideration in Section 3.5.

3.3.4 Turbulent Velocities

We need to know a value for the typical fluctuating velocity u of the turbulent eddies in the cylinder. To obtain this, we must discuss flame propagation in the cylinder.

In a turbulent pre-mixed medium, a flame propagates (in first approximation) at the turbulent fluctuating velocity. In Figure 3.5, I have drawn a cartoon of a flame front propagating in a turbulent medium. The flame front propagates at its laminar flame speed (somewhat modified by the stretching of the flame front by the turbulence), but as it propagates it is carried backwards and forwards by the turbulence. To the extent that the turbulent fluctuating velocities are substantially

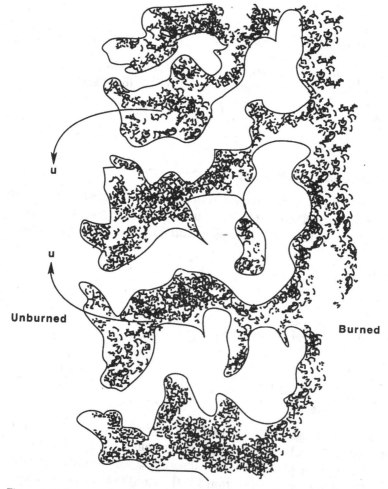

Figure 3.5. Cartoon of a flame propagating in a turbulent pre-mixed medium.

larger than the laminar flame speed, the rate at which the leading edge of the enflamed region (called the flame brush) propagates is roughly the turbulent velocity.
The enflamed region would thicken as it propagates, but the material (well) behind
the flame front has been consumed, and will not support a flame.

Why is this relevant? Well, consider: to a first approximation, combustion
occupies the same crank angle at all engine speeds and conditions. That is not
precisely true, because it is necessary to advance and retard the spark depending
on engine speed and manifold vacuum, but these are relatively small adjustments.
To a very crude approximation, combustion occupies the same crank angle regard-
less of engine speed. This means that the effective velocity of flame propagation is
proportional to piston speed, since each (flame and piston) must cross a fixed distance in the same time. And this means that the turbulent fluctuating velocity is
proportional to piston speed. In fact, Heywood [47] concludes that the turbulence
intensity at TC in an open combustion chamber (i.e., no squish, no pre-combustion
chamber) without swirl is $u \approx 0.5 \overline{V}_p$.

In discussing the turbulence intensity in the cylinder, we must specify TC,
because the turbulence intensity varies considerably during the cycle. The incoming charge produces a jet through the intake valve, and this jet is turbulent.
In addition, the mean flow energy of the jet is converted into turbulence in the
cylinder. However, this turbulence decays, so that it is considerably less by the
time the piston reaches TC. I include here Figure 3.6, which shows the turbulence

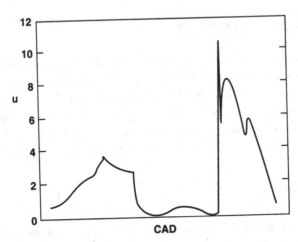

Figure 3.6. Turbulent velocity of unburned gas as a function of crank angle, generated by ESP for the CFR engine. It
begins when the intake valve closes. The first smooth rise
is due to the effects of compression. The abrupt change in
slope corresponds to ignition. The sharp peak is the end
of the burn. The smooth decrease is the effect of expansion (and decay); the sharp change in slope is caused by
the opening of the exhaust valve. The abrupt rise is caused
by the opening of the intake valve, producing a jet, which
intensifies and then diminishes as the instantaneous piston
speed first increases, and then decreases.

intensity generated by ESP for a CFR engine. The peak formed by the jet through the valve can be clearly seen.

3.3.5 Conclusions Regarding Temperatures

Knowing that $u \propto \bar{V}_p$, we can use Equation 3.12 to draw conclusions regarding the influence of various conditions and design decisions on part temperatures. For example, the temperature of an aluminum piston is considerably closer to the sink temperature than that of a cast iron piston. Heywood [47] says that cast-iron pistons run 40 K–80 K hotter than aluminum pistons, which is consistent. If an engine is scaled up geometrically, then $\frac{A_{conv}}{A_{cond}}$ will remain constant. Then, other things being equal, the ratio of temperature differences on the left in Equation 3.11 will be doubled in an engine (or cylinder) twice the linear dimensions. If the piston speed is doubled, the ratio on the left will be doubled. It becomes clear why the piston crowns melted in the VW engine. Also why the engine would drop a valve at highway speeds, rather than when idling. If the gas density is doubled (by supercharging) the ratio will be doubled. Thus, a supercharged engine with a small number of large cylinders, running at high speed, will be likely to have a cooling problem. It is clear that two small exhaust valves will run cooler than one large one (see Figure 2.7).

3.4
OVERALL HEAT TRANSFER

Data on heat transfer from local areas in the cylinder (e.g., exhaust valves, piston crowns) are relatively sparse and have a lot of scatter. However, one source of data is quite reliable: that on overall heat transfer, the total heat lost to the water jacket from the cylinder. This is reliable because it does not depend on local measurements, but rather on measurements of net heat lost to cooling water.

It is customary [94] to define an overall engine heat transfer coefficient by

$$h_e = \frac{\dot{Q}}{A_p(T_g - T_c)} \tag{3.13}$$

where $\dot{Q}$ is the heat lost from the gases per unit time, A_p is the piston area, T_g is the cycle mean gas temperature, and T_c is the mean coolant temperature. These are clearly quantities that are easy to measure, although not precisely the most meaningful ones. For example, we would expect that the heat transfer would depend on the total surface area in the cylinder; however, it is reasonable to expect that this will be roughly proportional to the piston area. Also, the heat lost should be proportional to the difference between the gas temperature and the wall temperature, but since most engines have similar construction, the temperature difference between the gas and the wall will usually be roughly the same fraction of the difference between the gas and the coolant.

Taylor [94] finds, from a great mass of data, that

$$\frac{h_e b}{k_g} = 10.4 \left[\frac{Gb}{\mu_g} \right]^{0.75} \tag{3.14}$$

where I am using notation slightly different from that of Taylor [94]. Here G has the dimensions of a density times a velocity; it is determined from the mass flux per unit time divided by the piston area:

$$G = \frac{\dot{m}_a}{A_p} \tag{3.15}$$

b is the bore. μ_g is the dynamic viscosity at the cycle mean gas temperature. k_g is the thermal conductivity at the cycle mean gas temperature.

We can see that $\frac{Gb}{\mu_g}$ is a Reynolds number, based on a mean (mass weighted) gas velocity and the bore. In the same way, since $h_e = S_t u \rho c_p$, we can write

$$\frac{h_e b}{k_g} = \frac{S_t \rho c_p u b}{k_g} = S_t \frac{ub}{\kappa_g} \tag{3.16}$$

where $\kappa_g = k_g / \rho c_p$ is the thermometric diffusivity. The last term is the Peclet number based on the turbulent fluctuating velocity and the bore. The Peclet number is like the Reynolds number, but uses the thermometric diffusivity κ_g instead of the kinematic viscosity. The thermometric diffusivity describes the molecular transfer of heat in a gas, just as the kinematic viscosity describes the molecular transfer of momentum. The Peclet number compares the transfer of heat by motion of the gas to the molecular transfer of heat, just as the Reynolds number compares the transfer of momentum by motion of the gas to the molecular transfer of momentum. The Peclet number is the Prandtl number times the Reynolds number. The Prandtl number is just the ratio of the kinematic viscosity to the thermometric diffusivity.

Cycle average gas temperature is a weak function of compression ratio and a stronger function of equivalence ratio. It is primarily determined by the heat released at combustion. However, although a higher compression ratio results in a higher peak temperature, it also results in a greater expansion of the exhaust gases, which produces a lower exhaust gas temperature, which dominates, and lowers the cycle average gas temperature. The effect is not large, however – a change in compression ratio from 8 to 10 reduces the cycle average gas temperature by only 26 K. It is a stronger function of equivalence ratio; however, it does not vary greatly near stoichiometric. For example (at an intermediate compression ratio), at an equivalence ratio of 0.8 it is 637 K, at 1.0 it is 680 K, and at 1.2 it is 677 K. It peaks at an equivalence ratio of about 1.1. For calculation purposes, we will use a compromise value typical of modern passenger car compression ratios, and corresponding to an equivalence ratio 1.1, 683 K. The corresponding values of the viscosity and conductivity are $\mu_g = 3.28 \times 10^{-5}$ kg/m sec, and $k_g = 5.11 \times 10^{-2}$ W/mK.

It is helpful to look at a numerical example. Let's consider the low compression CFR engine we met in Chapter 1, with bore $b = 8.26 \times 10^{-2}$ m, stroke $S = 1.14 \times 10^{-1}$ m, a volumetric efficiency of $\eta_v = 0.83$, at 1,700 rpm. Our cycle average gas temperature will be close enough for rough estimates, even at this low compression ratio. The inlet temperature $T_i = 322$ K, and the inlet pressure is 0.918 bar. At this temperature and pressure, the inlet density is $\rho_i = 0.992$ kg/m^3. The mass flux is given by

$$\dot{m}_a = V_d \rho_i \frac{N}{X} \eta_v$$

$$= (6.11 \times 10^{-4} \, \text{m}^3)(0.992 \, \text{kg/m}^3)\left(\frac{28.4}{2} \, \text{sec}^{-1}\right)(0.83)$$

$$= 7.14 \times 10^{-3} \, \text{kg/sec}$$

(3.17)

We also have the piston area $A_p = 5.36 \times 10^{-3}$ m^2, so that the Reynolds number is

$$R_g = \frac{\dot{m}_a b}{A_p \mu_g} = \frac{(7.14 \times 10^{-3} \, \text{kg/sec})(8.26 \times 10^{-2} \, \text{m})}{(5.36 \times 10^{-3} \, \text{m}^2)(3.28 \times 10^{-5} \, \text{kg/msec})} = 3.35 \times 10^3$$

(3.18)

From Equation 3.14 we immediately obtain $\frac{h_e b}{k_g} = 4.58 \times 10^3$. Using the values for k_g and b above, we obtain $h_e = 2.83 \times 10^3$ W/m^{2}K. Now, $T_g = 683$ K and $T_c = 85$ C $= 358$ K, so that $T_g - T_c = 325$ K. Also, $A_p = 5.36 \times 10^{-3}$ m^2. We obtain for the heat lost per unit time

$$\dot{Q} = h_e A_p (T_g - T_c)$$
$$= (2.83 \times 10^3 \, \text{W/m}^2\text{K})(5.36 \times 10^{-3} \, \text{m}^2)(325 \, \text{K})$$
$$= 4.93 \times 10^3 \, \text{W}$$

(3.19)

We are interested in the heat flux per unit (actual) area. Let's consider the same CFR engine, which has a simple cylindrical combustion chamber, of height h, the clearance height. The actual average area consists of that of the piston crown, and of the top of the combustion chamber, that of the walls of the clearance space, and one-half that of the walls of the cylinder. I say one-half the area of the cylinder walls, because this is a cycle average value. We will call this area A_T, and it will be given by

$$A_T = \frac{\pi b^2}{4} 2 + \pi bh + \frac{\pi bS}{2} = 1.97 \times 10^{-2} \, \text{m}^2$$

(3.20)

where we have used the equation

$$\frac{S + h}{h} = r = 4.92$$

(3.21)

to obtain $h = 2.91 \times 10^{-2}$ m. We then have for the heat flux per unit area

$$\dot{q} = \frac{\dot{Q}}{A_T} = \frac{4.96 \times 10^3 \, \text{W}}{1.97 \times 10^{-2} \, \text{m}^2} = 2.52 \times 10^5 \, \text{W/m}^2$$

(3.22)

Let us suppose that, on the average, the thickness of the walls separating the hot gases from the coolant is about $t = 6$ mm; using the figure we obtained earlier for $\gamma_{Fe} = 81$ W/mK, if T_s is the surface temperature, we can solve

$$T_s - T_c = \frac{\dot{q}t}{\gamma_{Fe}} = \frac{(2.52 \times 10^5 \text{ W/m}^2)(6 \times 10^{-3} \text{ m})}{81 \text{ W/mK}} = 18.7 \text{ K} \qquad (3.23)$$

We can also work out that

$$\dot{Q} = h_e^{true} A_T(T_g - T_s) = h_e A_p(T_g - T_c) \qquad (3.24)$$

from which we get that

$$h_e^{true} = h_e \frac{A_p}{A_T} \frac{T_g - T_c}{T_g - T_s} = h_e \frac{1}{3.68} \frac{327 \text{ K}}{308 \text{ K}} = 0.289 h_e = 8.16 \times 10^2 \text{ W/m}^2\text{K} \qquad (3.25)$$

From $N = 28.4$ rps we can calculate the average piston speed

$$\overline{V}_p = 2NS = 6.48 \text{ m/s} \qquad (3.26)$$

which gives us a turbulent fluctuating velocity at

$$u = 0.5\overline{V}_p = 3.24 \text{ m/s} \qquad (3.27)$$

We also have a density at TC, since $\rho_i = 0.992$,

$$\rho_g = r\rho_i = 4.92 \times 0.992 = 4.88 \text{ kg/m}^3 \qquad (3.28)$$

We can now calculate $\rho c_p u$:

$$\rho c_p u = \left(4.88 \frac{\text{kg}}{\text{m}^3}\right)\left(1 \times 10^3 \frac{\text{J}}{\text{kgK}}\right)\left(3.24 \frac{\text{m}}{\text{s}}\right) = 1.59 \times 10^4 \frac{\text{W}}{\text{m}^2\text{K}} \qquad (3.29)$$

which allows us to calculate the Stanton number

$$S_t = \frac{h_e^{true}}{\rho c_p u} = \frac{8.16 \times 10^2}{1.59 \times 10^4} = 5.13 \times 10^{-2} \qquad (3.30)$$

ESP is calibrated using Equation 3.14. That is, the Stanton numbers used in ESP are adjusted to make sure that the heat lost during the cycle is roughly that predicted by Equation 3.14. The Stanton numbers which produce this effect in ESP are 3.56×10^{-2}, which is certainly of the same order as that we obtained in Equation 3.30. The difference is attributable to the fact that ESP calculates separately the heat transfer during compression, during the burn separately in the burned and the unburned gas, and during expansion. ESP also takes account separately of the heat transfer from the gas jets through the inlet and exhaust valves. During each of these phases the turbulence levels are different. Considering the crudeness of our assumptions, we should be grateful that Equation 3.30 is as close to the ESP values as it is.

We can do one last calculation with these numbers, rough as they are. Suppose we want to know the temperature of the piston crown in our CFR engine, under these same conditions. To do this properly, we should solve a differential equation, looking at each section of piston crown between a radius r and a radius $r + dr$, adding the input from the gas, and the heat conducted into the section from the part toward the center of the crown, and subtracting the heat conducted to the part toward the edge of the piston. However, that is unnecessarily complicated when all we want is a rough estimate. We will use Equation 3.12. First, we must identify the various terms.

A_{conv} is clearly the area of the piston crown,

$$A_{conv} = \pi b^2/4$$

(3.31)

A_{cond} varies from the center of the piston crown to the edge. At any radius, it will be $2\pi rt$, where t is the thickness of the crown. Thus, it varies linearly, and we can use an average value:

$$A_{cond} = \frac{2\pi bt}{2} = \pi bt$$

(3.32)

The conduction distance is $L = b/2$. Thus, we can write

$$\frac{T_p - T_s}{T_g - T_p} = \frac{A_{conv}}{A_{cond}} \frac{h_e^{true} L}{\gamma_{Al}}$$

$$= \frac{b}{4t} \frac{b}{2} \frac{h_e^{true}}{\gamma_{Al}}$$

$$= 0.569$$

(3.33)

where we have used the value for $\gamma_{Al} = 204$ W/mK. We can now calculate T_p, the piston crown temperature, using $T_c = 358$ K and $T_g = 683$ K. We obtain $T_p = 476$ K = 203 C. Heywood [47] gives an average temperature of 260 C for a spark ignition engine at an estimated piston speed of 14.2 m/s. (Our crude prediction gives just a single figure for the whole crown, so we must compare with his average.) If we use Equation 3.12 again to adjust for the difference in piston speed, we can write

$$\frac{T_p^2 - T_c}{T_g - T_p^2} = \frac{\overline{V}_p^2}{\overline{V}_p^1} \frac{T_p^1 - T_c}{T_g - T_p^1}$$

(3.34)

where T_p^2 is the piston crown temperature at piston speed $\overline{V}_p^2$, and similarly for T_p^1 at $\overline{V}_p^1$. We had $V_p^1 = 6.46$ m/s, and $V_p^2 = 14.2$ m/s. This gives us $T_p^2 = 266$ C. That we are this close is probably an accident, because the calculation depends on assumptions regarding crown thickness, among other things. However, it does demonstrate that these numbers are not completely ridiculous.

Note that the heat transfer coefficient according to the measurements in Equation 3.14 is proportional to piston speed to the 0.75 power, rather than the 1.0 power as we have assumed in Equations 3.8, 3.9. This means that the Stanton number will vary as the 0.25 power of the Reynolds number. Fortunately, if we restrict our attention to production automobile engines we will not meet a wide range of Reynolds numbers, and 0.25 is a small exponent. However, it is a good idea to calculate the heat transfer coefficient using Equation 3.14, and then adjust it for small changes in conditions using Equation 3.12.

3.5
THE EXHAUST VALVE

The exhaust valve requires special treatment. Typical valve temperatures (see Figure 3.1) are in the neighborhood of 1,000 K, while cycle average gas temperatures are only about 700 K. If we were to use cycle average gas temperatures in Equation 3.12, we would find that the exhaust valve temperature is below the coolant temperature.

The problem is the fact that the valve lifts from its seat, changing the ratio of A_{conv}/A_{cond}, as well as the value of L, and exposing the valve to the hot exhaust gases. The heat gain during this portion of the cycle outweighs the heat lost during the remainder of the cycle. This is a very unsteady heat transfer problem. We can estimate a time constant for the heat transfer problem:

$$\mathcal{T} = \frac{\rho c_p V T L}{k_{solid} A_{cond} T} = \frac{LV}{\kappa_{solid} A_{cond}} \tag{3.35}$$

where V is the volume of the valve. That is: the time constant is the ratio of the internal energy content to the rate at which it is being conducted away, where the temperature gradient is being estimated as it was above, $\frac{\partial T}{\partial x} = \frac{T}{L}$. κ_{solid} is the thermometric diffusivity of the solid, $\kappa_{solid} = \left(\frac{k}{\rho c_p}\right)_{solid}$. If we take a typical valve in a CFR cylinder of diameter $b = 82.6 \times 10^{-3}$ m, with the valve diameter $D = 0.38b$, with a stem diameter of $0.25D$, we can make some reasonable assumptions about the geometry of the valve head and guess that the head volume is about 4.06×10^{-6} m^3. The distance the heat must travel is no more than about $L = 0.65D = 2.04 \times 10^{-2}$ m. The density $\rho = 7.87 \times 10^3$ kg/m^3, and the specific heat is $c_p = 0.47 \times 10^3$ J/kgK, while we may take $k_{steel} = 53$ W/mK as an average figure. This gives $\kappa_{steel} = 1.43 \times 10^{-5}$ m^2/s, and $A_{cond} = 4.84 \times 10^{-5}$ m^2. Finally, we have

$$\mathcal{T} = \frac{LV}{\kappa_{steel} A_{cond}} = \frac{2.04 \times 10^{-2} \cdot 4.06 \times 10^{-6}}{1.43 \times 10^{-5} \cdot 4.84 \times 10^{-5}} = 1.20 \times 10^2 \text{ sec} \tag{3.36}$$

That is, about two minutes. Roughly speaking, the head of the valve will cool to a temperature $1/e$ of its original temperature in a time $\mathcal{T} = 2$ min. This is not too surprising, when we consider what we know from common experience of how long it takes a piece of metal of this size to cool off. Our assumptions may be crude, but it is not possible to change the time constant $\mathcal{T}$ by as much as an order of magnitude by adjusting the assumptions. By contrast, the changes in the environment of the valve are taking place on a time scale between 10^{-1} sec and 10^{-2} sec, depending on engine speed. Hence, the physical picture is this: the valve temperature is essentially constant, while A_{cond}, A_{conv}, T_g, L all change very rapidly. This allows us to make an estimate of the heat transfer situation using the steady heat transfer equations we have derived.

Let's consider the product

$$(T_g - T_p)\frac{A_{conv}}{A_{cond}}L \tag{3.37}$$

which is proportional to the temperature difference between the valve and the cooling jacket, which is essentially constant. This product will have a different value in each part of the cycle. We can estimate these values, and average them together.

A typical exhaust valve has a stem about $0.25D$, where D is the head diameter. The stem enters the guide a distance about D below the head, and the valve seat has a height of about $0.07D$. When the valve is open, during the exhaust stroke, we can estimate A_{conv} as being the area of the top of the valve head, since the flow around the valve separates, and the heat transfer to the underside of the head is poor; there will be some heat transfer to the stem just before it enters the guide, where the flow reattaches, but the area is small, and we will ignore it:

$$A_{conv} = \frac{\pi D^2}{4} \tag{3.38}$$

On the other hand, A_{cond} is just the stem cross-sectional area. L will be less than D, since the heat is on average coming from the center of mass of the head; say, $L = 0.65D$. From this we get

$$\frac{A_{conv}}{A_{cond}} = \frac{\frac{\pi D^2}{4}}{\frac{1}{16}\frac{\pi D^2}{4}} = 16 \tag{3.39}$$

Taking $T_p = 1{,}020$ K, we also have $T_g - T_p = 1{,}300 - 1{,}020 = 280$ K. Thus, a product typical of the situation during the exhaust stroke will be

$$(T_g - T_p)\frac{A_{conv}}{A_{cond}}L \propto 280 \times 16 \times 0.65 = 2.91 \times 10^3 \text{ mK} \tag{3.40}$$

During the remainder of the cycle, the valve is closed (we are taking as a crude approximation that the exhaust valve is open for just the exhaust stroke; all we are looking for here is rough estimates). If the cycle average gas temperature is 700 K, and the exhaust gas temperature is 1,300 K for one quarter of the cycle, then the average gas temperature over the other three quarters of the cycle must be 500 K. The average temperature difference is $T_g - T_p = 500 - 1{,}020 = -520$ K, so we expect that on the average the valve is cooling during this part of the cycle.

Now $A_{conv} = \frac{\pi D^2}{4}$, and $A_{cond} = \pi D 0.07 D$, so that

$$\frac{A_{conv}}{A_{cond}} = \frac{\frac{\pi D^2}{4}}{0.07\pi D^2} = 3.46 \tag{3.41}$$

In addition, $L = D/2$, roughly. Thus, during this three quarters of the cycle our product will be

$$(T_g - T_p)\frac{A_{conv}}{A_{cond}} L \propto -520 \times 3.46 \times 0.5 = -900 \tag{3.42}$$

We can now obtain an average value of this product over the entire cycle:

$$\left[(T_g - T_p)\frac{A_{conv}}{A_{cond}} L\right]_{avg} = \frac{2910 - 3 \times 900}{4} = 52.5 \tag{3.43}$$

As I said above, the heat gained during the exhaust stroke outweighs the heat lost during the remainder of the cycle.

To show that these numbers work, the other side of the equation is

$$52.5 = (T_p - T_c)\frac{\gamma_{steel}}{h_e^{true}} \tag{3.44}$$

Taking $\gamma_{steel} = 53\,\text{W/mK}$, $h_e^{true} = 8.16 \times 10^2\,\text{W/m}^2\text{K}$ and $T_c = 358\,\text{K}$, we obtain $T_p = 1{,}160$, which is approximately correct.

We should also examine the heat transfer coefficient for the exhaust valve. The heat transfer situation when the valve is open is somewhat different from that we envisioned when we derived Equation 3.8. There is a jet of hot exhaust gases through the valve. The turbulent fluctuating velocity in the boundary layer on the valve will be about 1/30 of this jet velocity (see, for example, [96]). We can use the expression Equation 2.15. If we take $D = 0.38b$ from Figure 2.7, we can obtain $u/\overline{V}_p = 0.636$, which is relatively close to our estimate for the flow in the cylinder as a whole, which was $0.5\,\overline{V}_p$. We may thus expect that the heat transfer coefficient is about the same.

It is clear that this calculation (above) depends critically on the assumptions regarding the convection and conduction areas. I picked areas that would make the numbers come out right, but not the first time. The numbers I picked are reasonable but arguments could be found to increase or decrease any of these numbers somewhat, which would have a considerable effect on the equilibrium temperature. Such a calculation cannot be trusted to give reliable numbers without very careful measurements and/or detailed calculations of temperature distributions in the gas and solid. Even without this backup, however, the calculation does serve to indicate what is going on physically. It makes clear exactly why the cooling of the exhaust valve is such a critical matter.

ESP makes its calculations on an instantaneous basis rather than using averaged values, so that the Stanton number appropriate to the valve in ESP bears little relation to the number we have obtained here on an average basis.

3.6

CERAMIC COATINGS

In order to reduce the heat lost from the cylinder and increase the thermal efficiency, ceramic coatings have been introduced for the piston crown. We may
expect that these will reduce the temperature of the piston crown, in addition.
Realistically, we can also expect that adhesion will be a problem, since a ceramic and aluminum will in general have very different coefficients of thermal
expansion, and the piston is subject to large changes in temperature. This is an
old problem – jewelers have wanted to use enamel (which is also a ceramic) on
gold and silver, which (as tableware, for example) is ordinarily subject to much
smaller changes in temperature. Slowly, ways are being found of making the ce-
ramics stick. At present, these ceramic coatings are practical for racing, but not
for long-term street use.

In Figure 3.7 we have sketched a cross-section of a piston crown coated with a
ceramic, with the expected temperature variation. The ceramic has a poor coefficient of thermal conduction (compared to the aluminum of the piston crown), so
that even a thin coating (usually of order 1×10^{-4} m) of ceramic causes a non-
negligible fraction of the temperature difference to occur in the ceramic.

Since the temperature profiles are linear, the heat transfer problem is particularly simple. In addition, it is clear that the heat transfer rate through the ceramic
and through the piston crown will be the same. Let us call the heat transferred per
unit time per unit area *with the coating* $\dot{q}'$, and the heat transferred per unit area
and time *without the coating* $\dot{q}$. Let us denote by t_1 the thickness of the coating
and by t_2 the thickness of the piston crown, by γ_1 the thermal conductivity of the

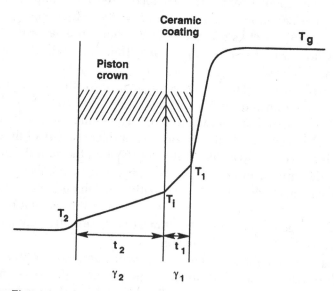

Figure 3.7. Variation of temperature through a piston crown with
a ceramic coating.

ceramic coating, and by γ_2 the thermal conductivity of the piston crown. Let us call the temperature at the surface of the ceramic coating T_1; we will use the same designation for the temperature at the surface of the piston crown *when there is no ceramic coating*. We will call the temperature on the bottom surface of the piston crown T_2, and the temperature on the interface between the ceramic and the piston crown T_i. Then we can say

$$\gamma_1 \frac{T_1 - T_i}{t_1} = \gamma_1 \left.\frac{\partial T}{\partial x}\right|_1 = \dot{q}' = \gamma_2 \left.\frac{\partial T}{\partial x}\right|_2 = \gamma_2 \frac{T_i - T_2}{t_2} = h_e^{true}(T_g - T_1) \qquad (3.45)$$

We may eliminate T_1 between one pair of equations, and T_i between the result and the remaining equation, to obtain

$$T_g - T_2 = \dot{q}' \left(\frac{t_2}{\gamma_2} + \frac{t_1}{\gamma_1} + \frac{1}{h_e^{true}} \right) \qquad (3.46)$$

If we designate $\Delta T = T_g - T_2$, we can write

$$\dot{q}' = \frac{\Delta T}{\frac{t_2}{\gamma_2} + \frac{t_1}{\gamma_1} + \frac{1}{h_e^{true}}} \qquad (3.47)$$

which we can compare with the uncoated piston

$$\dot{q} = \frac{\Delta T}{\frac{t_2}{\gamma_2} + \frac{1}{h_e^{true}}} \qquad (3.48)$$

The orders of magnitude of the coefficients are:

$$
\begin{aligned}
\gamma_1 &= 1\,\text{W/mK} \\
\gamma_2 &= 204\,\text{W/mK} \\
t_1 &= 10^{-4}\,\text{m} \\
t_2 &= 6 \times 10^{-3}\,\text{m} \\
h_e^{true} &= 8.16 \times 10^2\,\text{W/m}^2\text{K}
\end{aligned}
\qquad (3.49)
$$

so that when we form the ratios in Equations 3.47 and 3.48 we obtain

$$
\begin{aligned}
\frac{t_1}{\gamma_1} &= 10^{-4}\,\text{m}^2\text{K/W} \\[4pt]
\frac{t_2}{\gamma_2} &= 2.95 \times 10^{-5}\,\text{m}^2\text{K/W} \\[4pt]
\frac{1}{h_e^{true}} &= 1.23 \times 10^{-3}\,\text{m}^2\text{K/W}
\end{aligned}
\qquad (3.50)
$$

These are essentially impedances to the flow of heat. It is clear from this that the impedance in the gas dominates and essentially determines the heat transfer (even the ceramic has nearly an order of magnitude lower impedance), so that the ceramic coating will not have a large effect on the heat loss. In fact, the whole problem is probably easier to think about in terms of an electrical analogy. The ratio of the heat flux in the two cases is

$$\frac{\dot{q}'}{\dot{q}} = \frac{\frac{1}{h_e^{true}} + \frac{t_2}{\gamma_2}}{\frac{1}{h_e^{true}} + \frac{t_2}{\gamma_2} + \frac{t_1}{\gamma_1}}$$

$$= 0.926$$

$$(3.51)$$

The ceramic coating thus results in about a 7.4% reduction in heat loss. We are assuming that the temperature of the underside of the piston crown is kept constant (say, by an oil spray), and that the change in heat loss is not great enough to change the gas temperature appreciably.

The temperature drop through the piston crown with and without the ceramic coating can be obtained in the same way, by eliminating the heat flux between pairs of equations, or by the electrical analogy. If we designate by T_1^{wo} the temperature at the surface of the piston crown without the coating, then we obtain exactly the same expression:

$$\frac{T_i - T_2}{T_1^{wo} - T_2} = \frac{\frac{1}{h_e^{true}} + \frac{t_2}{\gamma_2}}{\frac{1}{h_e^{true}} + \frac{t_2}{\gamma_2} + \frac{t_1}{\gamma_1}}$$

$$= 0.926$$

$$(3.52)$$

What we are comparing is the temperature drop through the same thermal impedance at two different heat fluxes, so the ratio of the temperature drops must be the same as the ratio of the heat fluxes, and the temperature of the crown surface is only reduced by about 7.4%.

3.7
PROBLEMS

1. • Construct a crude steady state model that describes the heat transfer through a piston. The heat is transferred from the hot gases, through the piston crown, and out of the bottom of the crown to a spray of oil ($T_{oil} = 170°C$). Suppose the oil spray is generous enough to maintain the bottom surface at 170°C. Approximate the piston crown as a circular disk (diam = 80 mm, $t = 5$ mm) with a thermal conductivity of $\gamma_{Al} = 173$ W/mK. At max power, the average gas temperature in the combustion chamber is 450°C, and the top of the piston crown is 250°C. Sketch the problem, and set up an equation for the heat transfer of the system.

   • How much power does the gas in the cylinder lose by this path?

   • Add a 0.007 inch ceramic coating ($\gamma_c = 1.0$ W/mK) to the gas side of the piston crown. Obtain the temperature of the crown just under the ceramic coating. Assume the convective heat transfer coefficient h_e in the gas, T_{oil}, and T_{gas} remain constant. You will have to obtain h from solving the problem without the ceramic.

- Describe the important assumptions in the model.
- If you really wanted to find out whether the ceramic coating decreases the temperature enough to warrant a design change, then describe which assumptions would need further inquiry, and which would not.

2. Working with ESP, using no manifold and any convenient configuration, vary the "temperature of liner/piston/head heat transfer area" to 350 K, 400 K, and 450 K. Determine and compare the peak pressures (be sure to run more than one cycle to get convergence).

3. Our base engine has six cylinders, 3.4 L displacement, with $b = 73.7$ mm and $S = 133$ mm. The engine horsepower peaks at 4,500 rpm. The inlet valves have a diameter of 40 mm (permitted by a hemi head), and the cam produces an average flow coefficient of 0.5. There are two likely reasons why the engine would peak at this piston speed – which is it? (Justify your answer.) If the engine is redesigned to be square ($b = S$), with the same displacement, using the same valves,

- how much will the peak power increase (presuming that the *bmep* remains the same, and it peaks at the same piston speed)?
- The piston crown was running at 300°C. With the redesign, what will the piston crown temperature be (supposing the piston is geometrically similar)?

4. Marketing has suggested increasing the displacement of our 3.5 L six cyl engine to 4.0 L to meet the competition. Engineering suggests increasing the bore and keeping the stroke the same, so that the same crank can be used. The engine was square ($b = S$) and will now be over-square ($b > S$). The block is maintained at 85°C. Before the change, the piston crown was 6 mm thick, and ran at 300°C.

- If combustion temperatures and piston crown thickness remain the same, how much will piston crown temperature rise?
- To hold piston crown temperature the same, what should be the new piston crown thickness?

4
ENGINE FRICTION LOSSES

4.1

LUBRICATION

When two surfaces are in relative motion with a lubricant between them, the phenomenon will fall into one of three rough categories or types of lubrication. First, it is essential to realize that the surface of a smooth piece of metal is not smooth. Smooth is a relative term, and is entirely qualitative. Metal is given a finish which is specified in terms of the permissible roughness. A cylinder wall, for example, might have a $2\,\mu$m finish, meaning that the rms roughness is 2×10^{-6} m high. Bearing materials have similar finishes. Boundary lubrication (see Figure 4.1), the first type, involves metal-to-metal contact between the tops of the roughness elements (or *asperities*) of the two surfaces. This involves deformation and fracture of the surfaces and removal of bits of the surface. The second type is called mixed-film lubrication. The surfaces are separated by a slightly thicker film of lubricant now, and the metal-to-metal contact is only occasional. As the surfaces move farther apart and the lubricant film is thicker, the third type arrives, hydrodynamic lubrication. Now the surfaces never touch, and no wear takes place.

In an automobile engine, the main and connecting rod bearings are intended to operate in the hydrodynamic regime. However, when the engine is first started, if it has not run for several hours, say overnight, the oil film has probably been squeezed out of the bearing, and when first started it is operating in boundary lubrication. As the oil pressure rises, the oil film is replenished, and the lubrication becomes hydrodynamic. This is why bearing material is a lead-tin-antimony (and possibly arsenic) alloy, which is quite soft, so that the crankshaft will not be damaged by this metal-to-metal contact. It is almost fair to say that the life of an engine is measured by the number of times it has been started, and not by the number of miles it has driven.

The reason I say it is "almost fair" is the fact that there are other parts of the engine that are operating in boundary lubrication at various points during normal usage. Piston rings (see Figure 4.2 for example) operate in hydrodynamic lubrication through the mid-part of the stroke, maintaining an oil film thickness in the neighborhood of $5\,\mu$m [47]; at the top and bottom of the stroke, the rings lose much of their oil film, since the speed relative to the cylinder wall drops essentially to zero. Near BC, the film thickness drops to roughly $2\,\mu$m [47], and

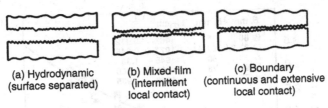

(a) Hydrodynamic
(surface separated)

(b) Mixed-film
(intermittent
local contact)

(c) Boundary
(continuous and extensive
local contact)

Figure 4.1. Three basic types of lubrication. The surfaces are greatly
magnified [54]. Copyright © 1983, 1991, by John Wiley & Sons, Inc.
Preprinted by permission of John Wiley & Sons Inc.

near TC, where the gas pressure forces the ring against the wall, the film thickness
drops to less than 1 μm. Certainly in the latter case, boundary lubrication results,
and it is near TC that most of the wear takes place. At low engine speeds, the
camshaft bearings probably operate in boundary lubrication, and the cam/lifter
interface certainly does. Hence, a certain amount of wear is unavoidable, and the
extent will depend on the duty cycle of the engine.

4.2
TOTAL ENGINE FRICTION

By driving a non-firing engine by an auxiliary motor, a measurement can be gotten
of the total friction losses. In the sections that follow we will consider the differences between a motoring engine and a firing engine, and what effect this might
have on the friction losses. We will also consider the attribution of the losses –
that is, how much of the losses can be attributed to the bearings, how much to the
piston rings, and so forth. Here, let us look at the total.

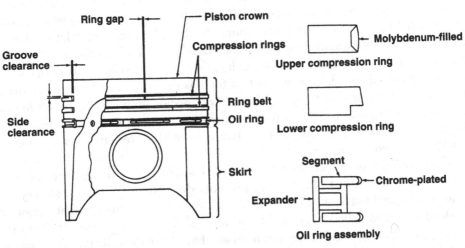

Figure 4.2. Construction and nomenclature of typical piston and ring assembly [61]. Reprinted
with permission from 820085 © 1982 Society of Automotive Engineers, Inc.

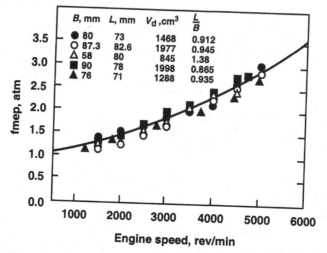

Figure 4.3. Friction mean effective pressure under motored conditions at wide open throttle for several four-cylinder spark-ignition engines [11]. Reproduced with kind permission of Institution of Mechanical Engineers, London, UK.

The friction mean effective pressure is defined just like the brake mean effective pressure:

$$P_f = fmep\, V_d \frac{N}{X}$$

$$(4.1)$$

where P_f is the power expended on friction.

In Figure 4.3 we see measurements of total friction mean effective pressure for several four-cylinder spark-ignition engines. Friction in an engine is due partly to the piston rings and piston skirts rubbing against the cylinder walls, and partly to pumping losses; we would expect both these losses to scale with piston speed. We will see later that this friction amounts to about half the total friction. However, the other half (crankshaft and seals, water pump and alternator, oil pump and valve train) does not scale particularly with piston speed, but rather with engine rotational speed. This means that there is no truly satisfactory single way to scale the total engine friction. The authors responsible for Figure 4.3 found that rotational speed produced a better correlation than piston speed for this data. Heywood [47] says that total friction data is normally presented versus engine rotational speed. However, the difference in scatter here is small; the data can be replotted against piston speed for an only slightly less satisfactory result. I show this in Figure 4.4. Taylor [94] presents data on *fmep* for two-stroke engines which includes a miniature engine of stroke 1.9 cm and a Sulzer diesel of stroke 1.2 m. The values of *fmep* fall on the same line when plotted against piston speed, which certainly seems to indicate that piston speed is the more useful variable.

Presenting the friction losses as *fmep* removes much of the effect of engine speed and displacement. As can be seen in Figure 4.3, engines of several different

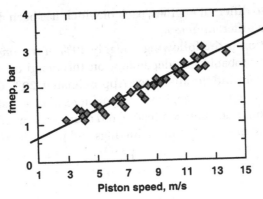

Figure 4.4. The data of Figure 4.3 replotted vs. piston speed.

displacements are clustered around the same curve. The range of displacements is not large, but at least for the displacements met with among passenger car engines, displacement appears to have been removed as a variable. This is also supported by the much larger range of displacements presented by Taylor [94]. There are evidently other variables, however. The relatively small scatter of Figure 4.3 is a little misleading. Taylor [94] gives data for the U.S. passenger car fleet for 1938; at a piston speed of 5.33 m/s the average *fmep* = 1.7 bar ±23%, which is a spread from 1.3 bar to 2.08 bar.

But, what about engines of different *bmep*? If we consider an engine of *bmep* = 0.75 MPa and one of *bmep* = 1.1 MPa, should we expect that the *fmep* will scale with the *bmep*, so that it is a constant percentage? An increased *bmep* will increase the pressure on the piston rings, causing greater friction, and will also cause greater side force on the piston skirt, causing greater friction. However, we will see below that this is not a large effect, the difference between no pressure and full pressure resulting in only a 10% change in *fmep*. An increase in *bmep* will certainly not affect the power spent on the alternator, but it will probably have an influence on the power expended on the oil pump and water pump because these will have to be larger for an engine of the same displacement but greater output. The water pump, at least, will have to have a capacity proportional to the output, so that the power required will be roughly proportional to the *bmep*, and probably the same is true of the oil pump. The connecting rod and main bearings will have to have bearing areas proportional to the *bmep*, and thus the power consumed will be proportional also. The power expended on the valve train will not necessarily increase, depending on what has been done to the valve timing, and what sort of valve gear is used.

Taylor [94] considers this question, but in the context of proposed changes in an existing engine. One could imagine, for example, increasing volumetric efficiency by a tuned intake manifold, increasing the *bmep*, but leaving the engine otherwise unchanged. The increase in *bmep* will have a relatively small effect on the *fmep* (which we will discuss below). However, this does not address our question, which relates to two different engines, each of which has been designed from the beginning to have a different *bmep*.

My best suggestion is the following: roughly 29% of the *fmep* (see below for the percentages) is probably not dependent on the *bmep* of the engine – this includes the valve train and the alternator. Approximately 29% should be directly proportional to the *bmep* of the engine – this includes the crankshaft bearings and seals, water pump, oil pump and connecting rod bearings. Roughly 42% is weakly dependent on the *bmep* – this includes the rings and piston skirts. If we lump the 42% with the 29%, and say as a crude approximation that 71% is not dependent on *bmep*, then we can write as a very rough approximation

$$fmep = fmep|_{82} \left[0.71 + 0.29 \frac{bmep}{bmep|_{82}} \right] \tag{4.2}$$

where $bmep|_{82}$ is the mean value in 1982, roughly 1 MPa, and $fmep|_{82}$ is the value in 1982 from Figure 4.3 or 4.4. If we are interested in an engine with a $bmep = 1.1$ MPa, and $fmep|_{82} = 1.75 \times 10^5$ Pa, this gives a new value of 1.80×10^5 Pa. If we are interested in an engine with a $bmep = 0.75$ MPa, the *fmep* now drops to 1.6×10^5 Pa.

4.3
ATTRIBUTION OF FRICTION LOSSES

Engine friction losses include the friction of the main and connecting rod bearings (and the wrist pins), the friction of the piston rings and the piston skirts on the cylinder walls, the power required to run the valve gear and the oil and water pumps, and the power expended in getting the gases into (and out of) the cylinder (the pumping loses).

In Figure 4.5 we show the motored friction mean effective pressure versus engine speed for a four-cylinder spark-ignition engine, from [47], [61]. The largest single contributor is the pistons, pins, rings and rods, and after that at low speeds the valve train (but at high speeds, the crankshaft and seals, followed by the valve train). The approximate percentages at mid-range are 12% for crankshaft and seals, 46% for pistons, rings, pins and rods, 23% for the valve train, 6% for the oil pump and 13% for the water pump and alternator.

These measurements are taken by progressively dismantling the engine, known as a breakdown test. That is, the complete engine is motored, then the water pump and alternator are removed and it is motored again. Then the oil pump is removed and it is motored again, and so forth. That is why pistons, rings, pins and rods are lumped together, whereas from a theoretical point of view it might have been nice to group the connecting rod bearings with the main bearings. Clearly, however, it is more convenient to remove the pistons, rings, pins and rods together (although it might have been interesting to remove the rings, and reassemble the engine without them).

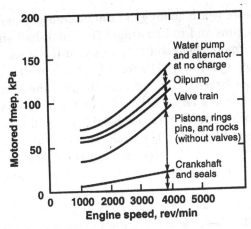

Figure 4.5. Motored friction mean effective pressure versus engine speed for a four-cylinder spark-ignition engine, from [47], [61]. Reprinted with permission from 820085 © 1982 Society of Automotive Engineers, Inc.

Motoring means that the engine is not running but is being turned over by an external agency (probably an electric motor), so that no combustion pressure is being generated in the cylinder. This means that the forces between the pistons and the cylinder walls are lower, and the bearing loads are smaller, than they would be in a running engine. The valve gear, however, is under essentially the same pressure. Tests have been made motoring an engine with steady pressure in the cylinders, and this does raise the *fmep* [94]. Considering the fraction of the cycle that the pressures from combustion are generated, Taylor [94] suggests for a four-stroke spark ignition engine that a steady pressure equal to 1/4 of the *imep* would give approximately the right correction to the *fmep*. Taking into account the sensitivity to steady pressure [94], this amounts approximately to the following formula:

$$fmep = fmep|_m + 1.65 \times 10^{-2} \; imep \tag{4.3}$$

where $fmep|_m$ is the motored friction mean effective pressure. This gives about a 10% correction to the total *fmep*, most of which is applied to the crankshaft and seals, and the pistons, rings, pins and rods.

In Figure 4.5 the power expended on each accessory is measured separately, and a corresponding mean effective pressure calculated.

According to Heywood [47], an approximate rule for estimating piston ring friction is that each compression ring contributes about 7 kPa to the *fmep* [71]. (This and the following figures in this paragraph are given at roughly 2,500 rpm or 6.4 m/s.) The oil rings, which have substantially higher ring tension, and operate under boundary lubrication, contribute about twice the friction of each compression ring [33]. This suggests that two compression rings and an oil ring (the standard complement) are worth an *fmep* of approximately 28 kPa, while pistons, rings, pins and rods are worth about 42 kPa at midrange, leaving perhaps 14 kPa for the piston skirts, pins and rod bearings. The crankshaft and seals contribute about 14 kPa to the *fmep*; Heywood [47] says that the seals are responsible for 20% of this [61], leaving about 11 kPa for the main bearings. The main bearings are usually somewhat larger in diameter, so that the shear in the oil is greater, causing greater stress, and a larger torque (since the lever arm is greater). In the Jaguar XK engine, for example, the main bearings would have an *fmep* roughly 2.7 times that of the rod bearings. If we take this as typical, this suggests about 4 kPa for the rod bearings, which suggests about 10 kPa for the skirts and pins. Of this, most is attributable to the skirts. We will talk more about the mechanics of bearings below.

The *fmep* of the valve train (see Figure 4.6) drive can vary as much as ±30%, depending on the spring force and valve train mass, and the critical speed. That is, the spring force is that required to hold the valve train on the cam at the critical speed, and hence will be proportional to the valve train mass, but the proportionality can vary depending on the critical speed – a lower critical speed

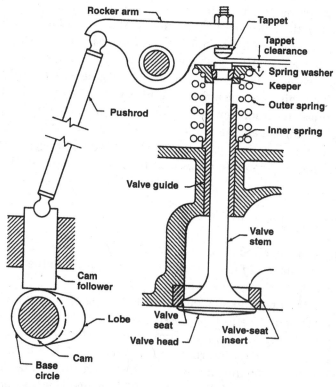

Figure 4.6. Poppet-valve nomenclature. An overhead valve mechanism with pushrod is shown [95]. Copyright © 1968 and new material © 1985 by The Massachusetts Institute of Technology.

means a weaker spring. Roller followers can result in reductions by as much
as 50% [47].

The water pump is typically less
than about 7 kPa at 1,500 rpm [47], [61],
and the alternator requires 7–10 kPa
[22], [47]. In Figure 4.5 our water pump
plus alternator consumed about 13
kPa at midrange, which approximately
agrees with those figures.

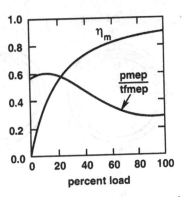

Figure 4.7. Mechanical efficiency η_m and ratio
of pumping *mep* to total friction *mep* as a function of load for a typical spark-ignition engine
at fixed speed [85]. Reprinted with permission
from 821576 © 1982 Society of Automotive Engineers, Inc.

As we have seen in Chapter 1, the
pumping losses vary from quite small at
WOT to very large at idle, and are quite
important for all partial throttle operation, which represents the majority of
normal driving. I reproduce here a figure from Heywood [47], Figure 4.7,
showing the ratio of pumping mean effective pressure to friction mean effective pressure. Pumping mean effective
pressure is defined exactly as for friction mean effective pressure, but using
the power expended on pumping, the
integral of the exhaust – intake loop of
the indicator card. It can be seen in Figure 4.7 that the *pmep* rises to nearly
60% of the total *fmep* at very low loads.

In Figure 4.8 I show curves of *pmep*
and rubbing *fmep*, or *rfmep*. This includes everything but the accessories
and pumping losses.

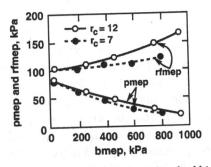

Figure 4.8. Pumping *mep* (*pmep*) and rubbing
friction *mep* (*rfmep*) as a function of load for
$r = 12$ and $r = 7$, four-cylinder SI engine with
$b = 95.3$ mm and $S = 114$ mm at 1,600 rpm [37].
Reprinted with permission from 580063 © 1958
Society of Automotive Engineers, Inc.

The *bmep* is a measure of load. This engine produces something like 950 kPa
with $r = 12$ with WOT, and when throttled down produces proportionately less.

4.4
HYDRODYNAMIC LUBRICATION

In Figure 4.9 I show a cartoon of a journal bearing at rest, in slow rotation, and
in full hydrodynamic lubrication. I am assuming that the bearing has been at rest
for some time in Figure 4.9a, so that the lubricant film has been squeezed out
of the bearing and the surfaces are touching. As the journal begins to rotate, the
frictional force causes it to be displaced to the right and upward. The lubricant
adheres to the surface of the journal, and to that of the bearing. (In fact, at normal

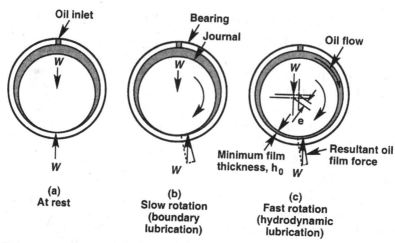

Figure 4.9. Journal bearing lubrication. The bearing clearances are greatly exaggerated [54]. Copyright © 1983, 1991, by John Wiley & Sons, Inc. Reprinted by permission of John Wiley & Sons Inc.

densities, all fluids adhere to the solid surfaces with which they are in contact, for complex statistical-mechanical reasons – this is called the no-slip condition.) The lubricant adhering to the surface is dragged into the contact zone. As more and more lubricant is dragged in, the lubricant in the converging zone before the contact zone builds up pressure which moves the journal to the left, and an equilibrium is established between the shear force, the pressure and the weight, with no metallic contact between the journal and the bearing surface. This is full hydrodynamic lubrication.

Note that it is not the pressure in the oil distribution system that causes the journal to float in the bearing, but the pressure generated by the wedging action caused by the shaft eccentricity (which is much larger than the system oil pressure). This wedging process is much the same physical phenomenon that is used in waterskiing. The forward speed of the skier corresponds to the shaft rotation; the angle of the skier's foot forms the wedge, and the water acts as the lubricant. The pressure generated in this way holds the skier up.

We can quantify this process a little. Let N be the rotational speed of the shaft (necessarily in rps.), which has dimensions T^{-1}. The bearing unit load P is defined as the load W divided by the bearing *projected area*, which is the journal diameter D times the bearing length L. P has dimensions M/LT^2. The dynamic viscosity is μ, which has dimensions M/LT. We can make a dimensionless number from these quantities: $\mu N/P$. It turns out that the coefficient of friction depends on this number in the hydrodynamic regime. We will justify this below by deriving this relationship from a mechanical model. For the moment, accept it as an experimental fact. The coefficient of friction is defined as $f = \tau/P$, where τ is the frictional surface force on the bearing per unit surface area. In Figure 4.10 I show what is called the Stribeck curve, relating f and $\mu N/P$. In boundary lubrication the coefficient of friction is not a function of relative speed, just as the coefficient

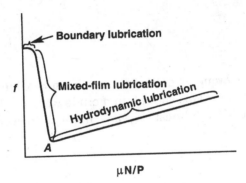

Figure 4.10. Coefficient of friction (and type of lubrication) vs. dimensionless variable $\mu N/P$ (Stribeck curve) [54]. Copyright © 1983, 1991, by John Wiley & Sons, Inc. Reprinted by permission of John Wiley & Sons Inc.

of friction of a block resting on a table is not a function of relative speed. It is evident that in hydrodynamic lubrication f is proportional to $\mu N/P$.

Let us derive the relation between f and $\mu N/P$. Let h be the average gap between the journal and the bearing, the gap which is filled with lubricant. Although the journal is located eccentrically (in order to support the load), the eccentricity is relatively small (we will comment on the eccentric case below). The velocity profile of the lubricant in the gap is very close to linear, making the velocity gradient essentially constant. The surface velocity of the journal is $r2\pi N$, where r is the radius of the journal. The bearing is not moving. The surface stress in a fluid in simple shear (like the lubricant in the gap) is given by

$$\tau = \mu\frac{\partial U}{\partial y} = \mu\frac{r2\pi N}{h} \tag{4.4}$$

With this, we can form the coefficient of friction:

$$f = \frac{\mu r2\pi N}{hP} = 2\pi\frac{r}{h}\frac{\mu N}{P} \tag{4.5}$$

We see that there is a multiplier which depends on the geometry of the bearing.

4.5
MECHANICAL EFFICIENCY

We have seen that in boundary lubrication, stress is proportional to load, and is not a function of speed. Because torque is proportional to stress times bearing surface area and radius, frictional torque will also not be a function of speed in boundary lubrication. On the other hand, in hydrodynamic lubrication, stress is proportional to speed, so that frictional torque will be also. We can write approximately

$$T_f = a' + b'N \tag{4.6}$$

where T_f is frictional torque, a' corresponds to boundary lubrication, and b' to hydrodynamic lubrication. This should, in a crude way, also bridge the gap between them, the mixed-film lubrication. Now, frictional power, $P_f = 2\pi N T_f$, so that

$$P_f = 2\pi a'N + 2\pi b'N^2 \tag{4.7}$$

We can also approximate the behavior of P_f by $P_f = AN^\alpha$, $1 \le \alpha \le 2$, where A is a constant and $\alpha = 1$ at low speed, and $\alpha = 2$ at high speed.

Now, the mechanical efficiency is defined as

$$\eta_m = \frac{P_i - P_f}{P_i} = 1 - \frac{P_f}{P_i} = 1 - \frac{fmep}{imep} \tag{4.8}$$

where P_i is the indicated power. At low engine speed, both the indicated power and the frictional power are proportional to N, say $P_i = \kappa_2 N$, $P_f = \kappa_1 N$, so that we have

$$\eta_m = 1 - \frac{\kappa_1 N}{\kappa_2 N} \approx 0.85 \tag{4.9}$$

for a typical engine; κ_1 is a constant of proportionality.

At high engine speed, P_i does not rise as fast as N, because of the fall-off of η_v. We can write $P_i = BN^\beta$, $0 < \beta < 1$. At the same time, the frictional power $P_f = AN^\alpha$, $1 < \alpha < 2$, and a lot closer to 2 than 1. Hence, η_m becomes

$$\eta_m = 1 - \frac{AN^\alpha}{BN^\beta} = 1 - \frac{A}{B}N^{\alpha-\beta} \tag{4.10}$$

where $0 < \alpha - \beta < 2$. We are subtracting something that is growing as the engine speed rises, and hence η_m goes to zero. In Figure 4.11 I show the mechanical efficiency for a CFR engine. I have taken the intake valve diameter $D = 0.44b$, and $C_i = 0.363$. I have supposed that η_v is constant at 0.83 until $Z = 0.6$, and then falls as $\eta_v = 1.08 - 0.442Z$, and I have taken $imep = \eta_v \times 1.16$ MPa. I have taken $fmep = 0.47 + 0.178\bar{V}_p$, the best fit to the data in Figure 4.4.

Some CFR engines have a balancing arrangement that contributes a large, fixed frictional load. As a result, the values of η_m for these engines are quite low. The engine referred to in Figure 4.11 is not of this type, however, so that we can expect that the curve in Figure 4.11 will also be quite typical of production engines; as we have seen, the $fmep$ rises steadily through the entire range of piston speed above about 3 m/s, so η_m drops continuously through the entire range. It begins to drop even faster when η_v begins to drop (when $Z \ge 0.6$, here about $\bar{V}_p \ge 14.5$). Typically, the curve reaches a value of 0.5–0.6 in the neighborhood of $\bar{V}_p = 20$ m/s.

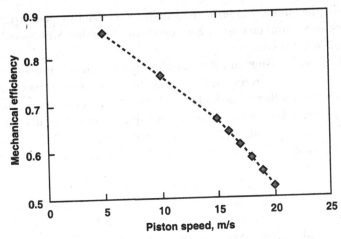

Figure 4.11. Mechanical efficiency vs. piston speed for a CFR engine.

4.6
INERTIAL LOADING

We have already seen that the friction coefficient is given by

$$f = \frac{\mu R 2\pi N}{hP} \tag{4.11}$$

The dependence of this expression on the gap h is not linear, and this has an interesting effect on the bearing drag when the gap fluctuates due to inertial loading.

The piston – connecting rod assembly is relatively massive, and its motion must change direction twice in each rotation. In fact, its motion is much more complicated than that, since the connecting rod big end is describing a circle, while the wrist pin is going back and forth on a line. However, we will not go into this complication here. As a very crude approximation, let us consider only the increased load near BC when the crank must bring the piston-connecting rod assembly to rest and start it upward, and the similar situation near TC. In addition, near TC every other time, combustion pressure will be added, which counteracts the inertial loading.

We can get a qualitative idea of the influence of this inertial loading. The inertial loading will push the bearing relative to the journal, so that the two are eccentric. Let us suppose that the displacement is ϵ. We can divide the bearing into quadrants: in one quadrant, the clearance is reduced to $h-\epsilon$, in the opposite quadrant, the clearance is increased to $h+\epsilon$, and in the other two quadrants the clearance is unchanged, at h. This will give a very crude estimate of the effect of this displacement. Then the net coefficient of friction for this bearing will be

$$\frac{1}{4}[f(h+\epsilon) + f(h-\epsilon) + 2f(h)] \approx \frac{\mu R 2\pi N}{Ph}\left(1 + \frac{\epsilon^2}{2h^2}\right) \tag{4.12}$$

Hence, any displacement increases the friction – the friction goes up more on the side of decreased clearance than it goes down on the side of increased clearance, so that there is a net increase.

Connecting rod bearings are designed so that the film thickness on the side of decreased clearance is reduced to no less than 2 μm under maximum inertial loading (this is the smallest film thickness that will maintain hydrodynamic lubrication with a reasonable surface smoothness specification). This means that ϵ is nearly equal to h. Equation 4.12 is valid only for $\epsilon/h \ll 1$; however, we can evaluate the left hand side of Equation 4.12 directly. If $\epsilon/h = 0.75$, we find a 64% increase in friction.

In order to reduce these reciprocating loads, which increase friction, the industry is considering (at the urging of the U.S. government) the possibility of tubular connecting rods and ceramic pistons, among other modifications. Ceramic pistons have the additional advantage of reducing heat loss substantially, increasing thermal efficiency. It is felt a net reduction of 10% in friction, resulting in a net gain in fuel economy of 2%, is possible by lightweight pistons, lightweight connecting rods, low tension rings (we will talk about these in the next section), the use of two rings per cylinder rather than three, and lightweight valve springs [76]. Even without exotic piston materials, it is possible to lighten the piston substantially. Aircraft experience (see [95]) has indicated that the piston height need not be more than 0.6b, and the skirt can be cut away in the areas that are not load-bearing.

4.7

THE PISTON RING

As we have seen, a single compression ring is worth about 7 kPa of *fmep* when motoring, and this is only increased by about 10% when corrected for compression/combustion pressure. This means that most of the *fmep* is due to the static tension in the ring, and only a small amount to the gas pressure. It is clearly desirable to think about low-tension rings, that would substantially reduce this static *fmep*. As noted above, the U.S. federal government has strongly recommended to the industry that such rings be considered.

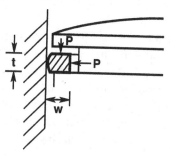

Figure 4.12. Cartoon of a compression ring being held against the cylinder wall by gas pressure.

The way in which gas pressure plays a role in holding the compression ring against the cylinder wall is illustrated in Figure 4.12. We can estimate the friction caused by this ring in the following way: the ring is pressed against the wall by the pressure acting on the vertical surface of the ring. The total area against which it presses is roughly πbt. The force then is $P\pi bt$, and the friction force is $\mu P\pi bt$, independent of velocity so long as the piston is moving. The average rate at which work is done by the friction force is

$$P_f = \mu P \pi \, bt \bar{V}_p \tag{4.13}$$

Now, the rings must be compressed to enter the cylinder, and this static spring force (which depends on the material and geometry of the ring) also presses the ring against the wall. When friction is measured by motoring, this static spring force is all that presses the ring against the cylinder wall. We may replace this static spring force by a fictitious pressure P_0, which we can compute from the contribution of the compression ring to motoring friction. We can write

$$P_f = \mu P_0 \pi \, bt \bar{V}_p = fmep \; V_d \frac{N}{X} = fmep \frac{\pi b^2}{4} \frac{\bar{V}_p}{2X} \tag{4.14}$$

from which we can compute

$$P_0 = fmep \frac{b}{t} \frac{1}{8\mu X} \tag{4.15}$$

If we consider that one compression ring results in $fmep = 7$ kPa, and if we take $b = 82.6 \times 10^{-3}$ m, $t = 1 \times 10^{-3}$ m, $\mu = 0.1$ for steel on steel with boundary lubrication, moving, we find that $P_0 = 3.61 \times 10^5$ Pa. This should be compared with the pressure at the end of the compression stroke, which depends on the compression ratio and the valve timing, but is typically between 0.9 MPa and 1.1 MPa for a modern engine. Thus, the pressure corresponding to the static spring tension is about 1/3 of the peak compression pressure.

There is a phenomenon known as ring float. In order to seal, the ring must remain pressed down against the bottom of the ring groove. If the ring lifts up, the gas escapes, and the pressure is lost. This is most serious at the end of the compression stroke near TC; during the power stroke, the pressures are much higher, and there is less possibility of this happening. As the piston approaches TC and leaves it, the upward acceleration is a maximum, tending to throw the ring upward by inertia. In addition, just after TC, when the piston has started down (but is still near the top, so that the acceleration has not decreased very much), the friction force on the ring is also upward. If we approximate the distance from TC by

$$x = \frac{S}{2}(1 - \cos \omega t)$$

$$V_p = \frac{S}{2}\omega \sin \omega t$$

$$a_p = \frac{S}{2}\omega^2 \cos \omega t \tag{4.16}$$

then the forces on the ring just after TC are approximately

$$\text{pressure force} = -P\pi \, bw$$

$$\text{inertia force} = \rho t \pi \, bw \frac{S}{2}\omega^2$$

$$\text{drag force} = \mu(P + P_0)\pi \, bt \tag{4.17}$$

If the forces are summed, we obtain

$$\omega^2 = \frac{P - \mu(P + P_0)\frac{t}{w}}{\rho t \frac{S}{2}}$$

(4.18)

We may evaluate this if we take $S = 1.14 \times 10^{-1}$ m, $t = 2 \times 10^{-3}$ m, $w = 5 \times 10^{-3}$ m, $\rho = 7.87 \times 10^3$ kg/m³, and $P = 1$ MPa, and $P_0 = 1.81 \times 10^5$ Pa. We obtain $\omega = 1.03 \times 10^3$, or $N = 9.84 \times 10^3$ rpm. This is too high to be of practical significance except on a racing engine. It does make clear that there is an advantage to a short stroke and thin rings, however. Long strokes and thick rings will make the situation worse. The situation is worse also for the second compression ring. If the pressure acting on the second ring is only 0.1 of that acting on the first ring, then the ring will float at 3,000 rpm.

This calculation is intended only to make clear what principles are involved. In practical situations, the pressure across the top compression ring is modified by leakage through the top and second rings, which is difficult to predict accurately, but which will surely reduce the net pressure difference, and lower the ω at which float will occur. For example, if the pressure *difference* holding down the top ring is cut in half by leakage and pressure retention by the second ring, but the top ring is pressed against the cylinder wall by the full pressure, this will cut the rpm for float to 6,780 rpm, which is now low enough to be a problem for passenger car engines.

4.8
PROBLEMS

1. As noted in Chapter 1, racing engines often are run with a single piston ring, since the leakage at low speed in not a problem for them. Suppose your sports car engine, with two compression rings and one oil ring has an *fmep* of

$$fmep = 9.76 \times 10^4 + 2.31 \times 10^4 \bar{V}_p.$$

(4.19)

Your base engine is a free-breathing four valve per cylinder hemihead. As a result, for WOT the *imep* $= 1.4$ MPa for $\bar{V}_p \le 25.1$ m/s, and thereafter follows

$$imep = 1.66 \times 10^6 - 1.55 \times 10^4 \bar{V}_p.$$

(4.20)

• Estimate the piston speed at which the mechanical efficiency of the base engine reaches 0.6. Use Equation 4.8 for the mechanical efficiency.

• For racing, you decide to remove one compression ring and the oil ring. Assume that each of the two compression rings is responsible for 7.7% of the total *fmep*, and that the oil ring is responsible for 15.3% of the total. Now estimate the piston speed at which the mechanical efficiency of the modified engine reaches 0.6.

2. In an older, low compression engine, peak combustion pressures are roughly 2 MPa above atmospheric. The piston rings are 3 mm thick, and are made of steel of density 7.6 gm/cc. The engine has a stroke of 185 mm. At what rpm will piston ring float begin? The coefficient of friction for lubricated steel on cast iron is 0.08.

3. Give four important effects that limit the piston speed in an engine. Explain each one in some detail (give equations, draw graphs, give numbers as/if appropriate).

FLOW IN THE CYLINDER

5.1

INTRODUCTION

The gas flow in the cylinder has a profound influence on the performance of the engine, whether intended or not. In the early days of the automobile, it was seldom intended, because relatively little was understood about turbulence, the primary player in this drama. Engines were designed largely by empiricism; from time to time a particularly successful design emerged, and its characteristics (to the extent that their relevance was recognized) were preserved in subsequent engines.

There are a couple of notable exceptions to this: In the first, between 1903 and 1907 H. R. Ricardo [45] (the father of the octane number), working at Cambridge University with Professor Bertram Hopkinson, did pioneering work on the effect of turbulence on combustion and heat transfer in the IC engine, particularly on the effect of the increased effective flame speed on knock, and on the possibility of stratified charge, among other things. This led, during the First World War, to great improvements in the design of tank engines (giving short flame travel and high turbulence levels, permitting higher compression ratios without knock), and after the war led to design modifications of the flat-head, or side-valve engine, which resulted in the same performance as the overhead valve engine, and which were generally adopted, and resulted in patents.

The other notable exception involved measurements made of the swirl and tumble produced by various inlet configurations, and the effect of the swirl and tumble on combustion [63], [86]. The flow measurements were made using high speed photography of goose down(!) in a glass cylinder. The combustion measurements used schlieren techniques. The authors determined the major mean flow and turbulent characteristics of swirl and tumble, and their effects on combustion (we will detail this later).

The work of Ricardo (above) and this pre-war work were largely forgotten, and had to be rediscovered during the 1980s and 1990s.

An example of an early engine designed largely on an empirical basis is the DOHC penta-head engine, which was developed for racing during the teens of the century, and was afterward extensively used in aircraft. It was designed principally to maximize the valve area, which keeps the Mach index as low as possible, although the Mach index was not understood at the time. It was known to be particularly successful, but it was not understood until recently (except for the work of Ricardo [45] and the NACA work [63], [86]) that the orientation of the inlet valves induces tumble, which is then broken up as the piston approaches TC, resulting in high turbulence levels, and high effective flame speeds. The central location of the spark plug also gives relatively short travel distances for the flame front, which Ricardo [45] understood. The combination of short distances and high effective speeds results in short burn times, which means that the compression ratio can be increased at the same octane number without knock, one of Ricardo's basic findings [45].

In modern engines, short burn times can be taken advantage of in other ways. For example, higher exhaust gas recirculation rates (EGR) could be tolerated, resulting in higher efficiency and lower NO_x production. This would be preferable to increasing the compression ratio in a modern engine, since an increased compression ratio could raise the unburned hydrocarbon emissions, because the crevice volume would become a larger fraction of the total volume. We will discuss all these possibilities in due course.

Now, engines are designed in an attempt to consciously bring about some of these effects. Unfortunately, our ability to apply existing limited understanding of fluid mechanics and turbulence to the complicated situation in the cylinder is still fairly rudimentary, and while the process is more rational and less empirical than it used to be, it still has some way to go.

Also, engine designers are trying to do rather ambitious things. These come under the general heading of flow management, with the goal of reducing brake-specific fuel consumption and emissions. There is a complex interaction among the flow management, the characteristics of the catalyst, the emissions, and the fuel consumption.

For example, the lean burn engine burns a homogeneous mixture of perhaps 24:1. Such a mixture is relatively difficult to burn, and requires a high turbulence level. Exhaust gas recirculation (EGR) is another technique to meet the same goal. EGR keeps approximately the same proportions, but replaces some of the excess air with recirculated exhaust gas, which has the advantage that a 3-way catalyst remains effective (current catalysts are not effective when the gases contain an excess of oxygen). Again, high turbulence levels are required to obtain reliable combustion.

The stratified charge engine attempts to segregate the incoming fuel vapor so that it does *not* mix with all the air in the cylinder, so that the engine will run at an effective air:fuel ratio of perhaps 50:1, while the fuel is confined to a small region of the cylinder volume where the air:fuel ratio is only 15:1, which will ignite and burn.

Stratified charge engines now are direct injection engines; that is, they inject the fuel spray at high pressure directly into the cylinder, using various strategies to keep the fuel spray from mixing with the entire contents of the cylinder. In the early development (say, twenty years ago) of the stratified charge engine, various techniques (other than direct injection) were used to segregate the charge, and they were not very effective. We will discuss some of these techniques later. Such engines did not achieve air:fuel ratios much above those of the lean burn engines, and are now thought to have been probably about as homogeneous as a poorly managed homogeneous charge engine. Terms like "nearly homogeneous," or "weakly stratified" might better be used to describe these engines. The suggestion here is, that "homogeneous" charge engines, unless considerable care be taken to induce homogeneity, are not particularly homogeneous.

Lean burn results in a lower combustion temperature overall, producing fewer oxides of nitrogen. Because the mixture is approximately stoichiometric or slightly rich in the fuel cloud in the stratified charge engine, the oxides of nitrogen are no lower than those of the lean burn engine, although the air:fuel ratio may be much higher. However, combustion quality is poorer than in stoichiometric homogeneous engines (due to slower burn and lower temperature), and as a result the emitted hydrocarbons are higher. CO is lower for the lean burn engine, but rises again for the stratified charge engine, due to the poor combustion. Moreover, lower levels of NO_x out of the engine do not necessarily result in lower levels of NO_x out of the tailpipe, because the conversion efficiency of the catalyst is low in the excess air of a lean-burn or stratified charge system. By comparison, EGR offers fuel economy and engine-out NO_x benefits comparable to a mildly lean combustion system, but allows the use of a three-way catalyst for oxidation of hydrocarbons and CO, and reduction of NO_x. This is also a cost-effective approach. Note that, for $150 Ryobi offers a 4-stroke-cycle gasoline weed-whacker with EGR – Los Angeles County take note! (Los Angeles County recently passed an ordinance outlawing weed-whackers because of their high emissions level.)

For at least the past fifteen years, the motivation for lean stratified charge engines in the automotive industry has been improved fuel economy. Not only are HC and CO emissions higher, but one has to start worrying about particulates, as in a diesel, because of the poor combustion. The few lean-burn engines that currently meet Federal emissions standards operate lean over only a very small portion of their speed-load map. None of them meets U.S. 50-state standards.

Management of flow in the engine cyclinder to bring about any of these effects is very tricky to do, as we will see, and the fluid mechanical details of how it works are poorly understood.

5.2

PHASES OF THE FLOW

The flow in the cylinder can be divided into several distinct phases. The flow into the cylinder through the inlet valve or valves (forming a jet) does two things: first, the geometrical configuration of the inlet ports and the valves, and their opening schedule creates organized motions in the cylinder, known as swirl (about the cylinder axis) and tumble (orthogonal to the cylinder axis); second, the jet itself is turbulent, and in addition much of the directed (non-turbulent) energy in the jet is converted to turbulence, resulting in a very high turbulence level during the inlet stroke.

During the second half of the inlet stroke much of this turbulence decays, which is to say that the intensity decreases markedly, both because the source (the jet) is coming to an end, and due to the effects of viscosity. The organized motions are carrying fuel vapor and droplets, residual gases, and all the contents of the cylinder, down the sides and across the bottom and up the other side, and round and round (depending on how much swirl and tumble there is). In addition, the turbulence is spreading itself, organized momentum, fuel, residual gases, in fact anything that can be transported, throughout the cylinder, trying to make everything as uniform as possible. This is what turbulence does best (transporting things), and it does it many orders of magnitude better than molecular transport.

During the compression stroke, the increase of density and the changes in length scales (due to the change in geometry of the charge as it is compressed) have the effect of amplifying the turbulence which remained from the inlet jet, although the viscous decay and turbulent transport continue. In addition, the swirl and tumble are affected by the same phenomenon.

If the piston and combustion chamber have been designed to produce squish (that is: if the piston crown approaches very close to some part of the combustion chamber roof), then this will have two effects: first, the fluid squeezed out of the squish clearance volume will produce organized motions, most of which will break up into turbulence; and second, the change in geometry due to the squish will have dynamical effects on the organized tumble and swirl.

Near TC, some of the organized motions may find they have insufficient room to maintain their form, and they will break up into turbulence, increasing the turbulence level. By this time conditions in the cylinder have become crudely homogeneous, due to the transporting effect of the turbulence and the organized motions, unless a concerted effort has been made by the designer to avoid this. Bear in mind, that the time available for transport (measured by the time scale of the turbulence itself) is not large, so the uniformity achieved can only be crude.

During combustion the turbulence level rises somewhat. Then, during the power stroke, the geometrical changes result in a strong attenuation of the turbulence, and any organized motion that has survived. This, combined with the viscous decay, results in the turbulence being sharply suppressed, so that by the time the exhaust valve opens, there is virtually nothing left. Very little turbulence is generated during the exhaust stroke.

We will examine these various phases one by one.

5.3
AVERAGING

We need to talk first about averaging, since turbulence is defined by fluctuations about an average velocity.

If we were dealing with a process that was not changing statistically from instant to instant (even though the instantaneous values were chaotic) such as the wind speed at midday on a day without severe changes in the weather, then we could use a time average, integrating over a time long compared to a typical time over which the values fluctuate. Such a situation is called stationary. Unfortunately, the situation in an engine cylinder is not stationary. Successive cycles are similar, but within each cycle the situation is statistically quite different from instant to instant.

In such a situation, it is useful to use a phase average. Imagine measuring, say, the circumferential velocity U at a particular location in the combustion chamber. This will be a function of crank angle θ and of the particular cycle in which we measured it, say

$$U(\theta, i)$$

(5.1)

where i indicates the number of the cycle in which it was measured. Then the phase averaged value is defined as

$$\overline{U}(\theta) = \frac{1}{N} \sum_{i=1}^{N} U(\theta, i)$$

(5.2)

where N is the number of measurements available. There are a number of interesting questions which we could address, such as, how many individual values do we need to include in Equation 5.2 to arrive at a stable value? This and many other questions are covered in [96], and need not concern us here.

Now that we have a convenient definition of an average value, which is a function only of crank angle, we can define the turbulent fluctuating velocity as the difference between the instantaneous value and the average value:

$$u(\theta, i) = U(\theta, i) - \overline{U}(\theta)$$

(5.3)

Statistics of $u(\theta, i)$ can now be defined as (for example)

$$\overline{u^2}(\theta) = \frac{1}{N} \sum_{i=1}^{N} u^2(\theta, i)$$

(5.4)

the mean square value.

Note that the mean values defined in this way are functions only of the crank angle. There is another way to define an average in this situation. Consider the time scale of the turbulence. For example, we will find below that the turbulence during the middle third of the intake stroke has an intensity (the root-mean-square value, from Equation 5.4) of roughly $10 \overline{V}_p$, where $\overline{V}_p$ is the average piston speed, and a length scale of roughly $b/6$ [98], where b is the bore. This means it has a time scale of roughly $\frac{b}{60\overline{V}_p}$. This time scale is $1/60$ the time it takes the piston to complete the intake stroke (presuming that the engine is nearly square, $b \approx S$). Now, what can we do with this? It tells us that there is a time, say $S/\sqrt{60}\overline{V}_p$, that is about $\sqrt{60} \approx 8$ times larger than the times typical of the turbulent

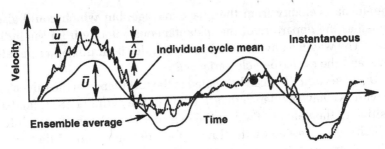

Figure 5.1. Cartoon of velocity variation with crank angle at a fixed location in the cylinder during two consecutive cycles of an engine. Dots indicate measurements of instantaneous velocity at the same crank angle. The phase averaged velocity obtained by averaging over a large number of such measurements is shown as a solid smooth line. The individual cycle mean is shown as a dotted line. I have shown a considerable difference between the individual cycle mean and the phase average. See [81]. Reproduced by permission of Plenum Press and W. C. Reynolds.

fluctuations, and yet shorter than (about $1/\sqrt{60} \approx 1/8$) the time of the intake stroke.

$$\frac{S}{60\overline{V}_p} \ll \sqrt{60}\frac{S}{60\overline{V}_p} = \frac{S}{\sqrt{60}\,\overline{V}_P} \ll \frac{S}{\overline{V}_p} \tag{5.5}$$

where $\sqrt{60} \approx 8$. We can use what is called a moving average:

$$\langle U(\theta, i)\rangle = \frac{1}{\Theta}\int_{-\Theta/2}^{+\Theta/2} U(\theta + \psi)\,d\psi \tag{5.6}$$

where $\Theta = S/8\,\overline{V}_p$. We can call this the individual cycle mean. I include here a figure from [47], Figure 5.1, which shows a cartoon of such an instantaneous velocity, the phase average, and the individual cycle mean.

This question of whether there is an intermediate time scale, larger than the small scale of the turbulence, but shorter than the longer scale of the intake stroke, is the same question that arises in meteorology, whether there is an intermediate scale between turbulence and weather, that would permit a distinction between them. It also arises in viewing pictures in a newspaper. The distance between the Benday dots that make up the picture is the fine scale, and the scale of the detail in the picture is the large scale. For the picture to be interpretable by our eye, a scale must exist that is large compared to the Benday dot scale and small compared to the scale of the detail. The situation is only unambiguous when there is a wide separation between the scales. There are many interesting statistical questions that can be asked about this situation – some of them were discussed in [66].

Heywood [47] says there is considerable debate in the literature over whether the cycle-to-cycle variation in the individual cycle mean is a physically distinct phenomenon from the turbulence, which we have defined as the departure of the instantaneous velocity from the phase average, but which could also have been defined as the departure of the instantaneous velocity from the individual cycle mean. The way we have defined it, it includes both this other definition of turbulence, and the cycle-to-cycle variation.

Cycle-to-cycle variability can be related to the turbulence in the inlet manifold. To begin with, we have evidence that the flow in the cylinder at TC (and after) is very sensitive to the velocity field in the cylinder at the moment the inlet valve closes, which is determined by the flow in the inlet manifold while the cylinder was filling. Exponential dependence on small differences in initial conditions is a characteristic of turbulence [49]. In the case of flow in the cylinder, the dependence on the initial conditions is known both from computational evidence and from experimental evidence. In [80] the authors examine the flow in a single cylinder engine with a pancake-shaped combustion chamber, with an inlet and exhaust manifold. The inlet valve is shrouded, to produce high swirl; such a configuration also has relatively little cyclic variability of the large scales. The authors calculated the mean flow structure, modeling the small scales of the turbulence. They found that the swirl ratio, and the phasing of the swirl structure 90 CAD ATC was very sensitive to the initial velocity field at inlet valve closing.

In [79], the author made measurements of the flow in the same engine, this time without the shroud on the inlet valve, so that the flow is undirected. The measurements were made by introducing very small particles that follow the flow, and photographing them at closely spaced times, obtaining velocity vectors from the differences in position. This is known as Particle Image Velocimetry, of PIV. In Figure 5.2 I show this flow. From Figure 5.2 it is clear that there is substantial variation in the large scale structure at TC. This variation in large scale structure will affect the flame propagation and heat release, and must consequently be responsible for the cycle-to-cycle variability.

In [43] the author carried out preliminary calculations of the unsteady flow in this same engine, including the inlet and exhaust manifolds, resolving only the largest scales (modeling the smaller scales). Again, he found large variability of the large scale structures, completely consistent with the PIV measurements and the mean flow calculations.

The flow is deterministic, and the flow at TC is entirely determined from (and very sensitive to) the flow at the inlet valve closing. Since this is entirely determined by the flow in the inlet manifold during filling, it must be this flow that is responsible for the cycle-to-cycle variability. The flow in the inlet manifold is very compressible as we have seen, and large amplitude sound waves are present in the manifold. It might at first be thought that these were responsible for the fluctuations in manifold flow. Measurements of the pressure fluctuations in the manifold of a 2.2 L Renault J7T (four cylinders, with two valves per cylinder) at several engine speeds [14], however, show that the pressure fluctuations (and hence the sound waves) are quite periodic; the very small differences from cycle to cycle are below the resolution of the pressure transducer.

We have to conclude that the differences in the initial velocity field in the cylinder when the inlet valve closes are attributable to the turbulent velocity

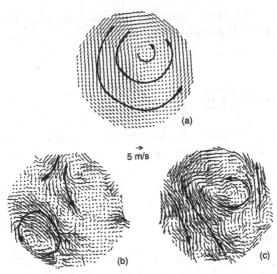

Figure 5.2. PIV measurements of the flow in the TCC (transparent combustion chamber) engine, from [79]. The engine has a pancake-shaped chamber, 12.2 mm clearance at TC. The measurements are at the mid-plane, and show a 70 mm circle in a 92 mm bore. The measurements are at 1,200 rpm and 40 kPa mean atmospheric pressure (i.e., at partial throttle). (a) shows the 90 cycle mean flow; (b) and (c) show two individual cycles.

fluctuations in the inlet manifold. These may have an indirect influence also, since the velocity fluctuations affect the pressure drop across the inlet valve during filling, and may have a small effect on the sound waves in the manifold. The relatively small differences at inlet valve closing attributable to the manifold turbulence are enough to cause considerable variability at TC, due to the extreme sensitivity of the turbulence to initial conditions.

Note that the various kinds of average (the phase average, and the moving average, which could also be applied in *space* within the engine cylinder) are not statistically equivalent. It is necessary in any discussion to be clear about which average is being discussed. Since the cycle-to-cycle variation, and the small scale turbulence in the cylinder appear to be physically the same, differing only in scale, it probably makes sense to lump them together, because the cycle-to-cycle variations are also responsible for transport in the cylinder. *In the rest of this book, I will use the definition of turbulence as the departure from the phase average.*

This does not mean, however, that the cycle-to-cycle variation should be ignored; it is responsible for unevenness in the running of the engine, and is generally undesirable for that reason. In addition, if there is cycle-to-cycle variation, some of the cycles will be of lower efficiency, and this will affect the average efficiency. I would rather describe the situation this way, however: that it is desirable to suppress as much as possible the largest scale variability of the flow entering and in the cylinder (which will be responsible for perceptable unevenness and fluctuation in efficiency, the longer, the more perceptible). This says the same thing, but the emphasis is a little different.

5.4

A WORD ABOUT TURBULENCE

The reader will need to understand just a little about turbulence for this discussion to make sense. It will not be necessary to consult a text on the subject, such as [96].

All fluid flow is either laminar or turbulent. The flow of pancake syrup is an example of laminar flow – smooth, steady and uncomplicated. When the kitchen faucet is turned on full and the flow leaves the faucet in an unsteady, rough, chaotic manner, filled with small eddies, that is turbulent flow. Most of the flow in the universe is turbulent.

Whether a flow is laminar or turbulent is usually (and certainly in our case) determined by a Reynolds number, which compares inertial to viscous forces. If U (m/s) is a typical velocity in the flow, L (m) a typical length, and ν (m²/s) the kinematic viscosity, a Reynolds number is defined as UL/ν. It is dimensionless. In any given flow, there are usually several Reynolds numbers that can be defined, depending on what velocity and length scales are used, but they are all related. The flow of pancake syrup is dominated by viscous forces, while the water flow is not. The Reynolds number of the pancake syrup is low, while the Reynolds number of the water flow is high. If the Reynolds number is large, the flow will probably be turbulent.

As the Reynolds number increases, a laminar flow becomes unstable. That means, that small disturbances, which would have been suppressed by viscosity when the Reynolds number was smaller, will instead be amplified, and the flow will become chaotic and unsteady. In this way, energy in the mean flow is transferred to the turbulence.

Turbulent flows contain a wide range of scales – that is, the velocity field contains motions of all sizes, from eddies that are essentially large enough to fill the space available, in our case the engine cylinder, down to eddies often substantially below a millimeter in size. The size of the largest eddy can be guessed by asking for the diameter of the largest sphere that will fit in the available space, since turbulent eddies are approximately the same size in all directions. Hence, at TC the largest eddy will be roughly the clearance height, while at BC it will be roughly the cylinder bore. The kinetic energy of these eddies varies: the largest eddies are relatively weak; as the size drops, the energy rises rapidly to a peak, and then falls continually down to the smallest eddies. The *most energetic* eddy, at the peak, which is responsible for most of the transport, is about 1/6 the size of the largest eddy – thus, 1/6 of the bore, or 1/6 of the clearance height [98].

Turbulence consumes energy. Mean flow energy is converted to large scale turbulence and those large scale motions are themselves unstable, and break down into smaller scales, and those in their turn break up into still smaller scales, until the smallest scale is reached. It is a curious experimental fact that the amount of energy consumed in a turbulent flow has nothing to do with the viscosity, but is determined entirely by the large turbulent scales. That is, the amount of energy taken from the mean flow is determined by the process of creating the largest turbulent eddies. This amount of energy is passed on to the next smaller eddies, and on to still smaller eddies, and is finally dissipated by viscosity at the smallest scales. The amount of energy passed from the mean flow to the largest eddies and on to eddies of progressively smaller size is called the dissipation, the energy consumed per unit mass of the flow, and is designated by ϵ. If u is a turbulent velocity scale, the root mean square fluctuating velocity (typical of the most energetic eddies), and ℓ is a length scale of the most energetic eddies, then it is an experimental fact that

$$\epsilon = \frac{u^3}{\ell} \tag{5.7}$$

You can understand this by writing $u^3/\ell = u^2(u/\ell)$. u^2 is proportional to the energy of the turbulence, and ℓ/u is proportional to the natural time scale of the most energetic turbulent eddies, which determines how fast the most energetic eddies lose energy to the next size eddies. From this you can see that u^3/ℓ is like du^2/dt.

The dynamical behavior of the smallest eddies is determined entirely by the rate at which they are fed energy, ϵ, and the size of the kinematic viscosity ν. The small eddies are where the mechanical energy is converted to heat, although the *rate* of conversion is determined by the large scales. If the viscosity is raised, ϵ does not change, but the scale at which the dissipation takes place increases. The smallest scale is called the Kolmogorov microscale, designated by η, and defined as

$$\eta = \left(\frac{\nu^3}{\epsilon}\right)^{1/4} \tag{5.8}$$

Smaller scales cannot exist because the viscous transport smears them out. In most flows the Kolmogorov microscale η is a few tenths of a millimeter. We will find that in an engine cylinder, it is of the order of 0.01 mm, or 10 μm.

As I noted above, in an engine cylinder, we expect the most energetic scale to be of the order of $h/6$ if the piston is near TC, or of the order of the $b/6$ if the piston is halfway down or more. Consider a cylinder halfway between TC and BC on the intake stroke, and suppose the engine is running at a piston speed of 5 m/s. Assume an inlet density of $\rho = 0.5$ bar, and suppose $S=b$. We will find in Section 5.5 that the turbulent velocity when the piston is in the middle of the intake stroke is about 10 times the mean piston speed, or perhaps 50 m/s here. We expect that $\ell \approx b/6$, and let us take $b = 82.6$ mm. This then gives us an $\epsilon \approx 9$ MW/kg, giving $\eta \approx 7.4 \times 10^{-6}$ m $= 7.4$ μm. As turbulent flows go, in the environment and the laboratory, this is extraordinarily small; η in the atmosphere or the ocean is about 1 mm, and in laboratory experiments not much smaller than a fraction of a millimeter.

The value of $\epsilon \approx 9$ MW/kg sounds enormous, but bear in mind that the mass in a cylinder is less than a gram, that this level of dissipation does not last long:

roughly $S/3\,\bar{V}_p \approx 5.5$ ms, and it results in a temperature rise of less than 50 K. For higher piston speeds ϵ rises very fast, since $\epsilon \propto \bar{V}_p^3$. η will fall slowly as $\eta \propto \bar{V}_p^{-3/4}$.

We can look at the cylinder at TC. Now ℓ is 1/6 of the clearance height, h. If $r = 8$, the clearance height is approximately $h = 11.8$ mm. Hence, $\ell = h/6 \approx 2$ mm. We have already seen in Section 3.3.4 that at TC, in the absence of swirl, tumble and squish, $u \approx \bar{V}_p/2$. The kinematic viscosity in a gas is proportional approximately to $\sqrt{T}/\rho$. With $r = 8$, the isentropic temperature at the end of compression has increased by a factor of 2.3, while the density has increased by a factor of 8, giving a value for the kinematic viscosity of $\nu = 2.84 \times 10^{-6}$ m^2/s. With $\bar{V}_p = 5$ m/s, this gives $\epsilon \approx 7.9$ kW/kg, and a value of $\eta \approx 7.34$ μm, still about the same size, and quite small by non-automotive standards.

Notice, if we use both our Equations 5.7 and 5.8, we can write

$$\frac{\eta}{\ell} = \left(\frac{\nu^3}{\epsilon\ell^4}\right)^{1/4} = \left(\frac{\nu^3}{u^3\ell^3}\right)^{1/4} = R_\ell^{-3/4} \tag{5.9}$$

where $R_\ell = u\ell/\nu$ is the Reynolds number of the turbulence, based on the turbulent fluctuating velocity and the length scale of the energy-containing turbulent eddies. Notice that this Reynolds number determines the ratio of the energy-containing and small scales in the turbulent flow; when $R_\ell \approx 1$, the energy-containing and small scales are approximately the same, and at this point the turbulence can no longer maintain itself. To exist, turbulence needs to have a range of scales; when the small scale and the energy-containing scale are the same, this means that viscosity acts directly on the energy-containing scales, and that makes the turbulence too dissipative, and it dies. At our two states, halfway down on the intake stroke and at TC, we have scale ratios of $\ell/\eta \approx 1.9 \times 10^3$ and 1.6×10^2, for turbulent Reynolds numbers respectively of $R_\ell \approx 2.3 \times 10^4$ and $R_\ell \approx 8.7 \times 10^2$.

In a gas, ν measures the molecular transport of momentum. $\nu \approx c\lambda$, where c is the root-mean-square molecular thermal velocity, and λ is the mean free path. That is, ν is proportional to the product of the velocity scale and the length scale of the process responsible for the property transport, the molecular motion. In a turbulent fluid, the turbulence transports properties, by simply carrying the fluid containing them to a new location. This transport process is nothing like (phyically or mathematically) the molecular transport; nevertheless, it is often convenient to parameterize it by what is called an eddy viscosity, ν_T. Just like the molecular transport, ν_T, the turbulent transport coefficient, is also proportional to the product of the length and velocity scales of the physical process responsible for the transport, the turbulence. Hence, $\nu_T \approx u\ell$. We can write

$$\frac{\nu_T}{\nu} = \frac{u\ell}{\nu} = R_\ell \tag{5.10}$$

so that, in our two representative situations (halfway down on the intake stroke, and at TC on the compression stroke) the turbulence is approximately 10^4 and 10^3 times as effective at transporting momentum (and everything else) as the molecular motion. Values of ν_T in the two situations are $\nu_T \approx 6.9 \times 10^{-1}$ m^2/s and $\nu_T \approx 1.4 \times 10^{-2}$ m^2/s, far larger than the molecular values.

During a time t, the turbulence will carry a property (momentum, for example) a distance L of roughly [96]

$$L \approx \sqrt{\frac{2}{3}v_T t} = \sqrt{\frac{2}{3}u\ell t} \qquad (5.11)$$

Consider the situation half-way down on the inlet stroke. If we measure t in terms of the time to complete the inlet stroke, say 1/3 of that time, when the turbulence is most intense, $t = S/3\bar{V}_p$, then

$$\frac{L}{b} \approx \sqrt{\frac{2}{3}u\frac{\ell}{b^2}t} \approx \sqrt{\frac{2}{9}\frac{u}{\bar{V}_p}\frac{\ell}{b}\frac{S}{b}} \approx 0.6 \qquad (5.12)$$

Thus, this intense turbulence is capable of transporting momentum across a substantial fraction of the engine cylinder in the time available. Note that it transports fuel vapor (or anything else) just as well as momentum, since the transport is physical – the fluid containing the momentum or fuel vapor is simply moved to another place, so the effectiveness of the transport does not depend on the nature of the thing being transported, and it works as well for all properties.

On the other hand, at TC, the turbulence is much weaker, and the time available is much shorter. The burn time corresponds to roughly $0.28 S/\bar{V}_p$, so that

$$\frac{L}{h} \approx \sqrt{\frac{2}{3}\frac{u}{\bar{V}_p}\frac{\ell}{h}\frac{S}{h}0.28} \approx 0.33 \qquad (5.13)$$

so only a fraction of the clearance height can be covered in the time available.

There are, of course, other factors involved, and these are extremely approximate values, intended only to give the flavor of what is going on. It is clear, however, that the turbulence available during the burn is marginal, considering that it is primarily responsible for the flame propagation. It is equally clear that it will be very difficult to keep the fuel vapor segregated during the inlet stroke, in the face of the aggressive transport properties of the turbulence available then.

ESP (see Chapter 8) assumes that the charge is homogeneous – that the aggressive transport during the intake stroke has resulted in uniformity. To calculate the turbulence level, it uses turbulence models that are (in some respects) a little more sophisticated than our crude approximations, and in others less so.

5.5
TURBULENCE INDUCED BY THE INLET JET

I will assume here that we are discussing a homogeneous charge engine, without significant swirl or tumble, which I will address later.

The flow through the inlet valve forms a hollow conical jet, roughly at the angle of the valve seat. We may expect that the Reynolds numbers will in general be high, since the viscosity of air at room temperature is relatively low. Using our estimate for the average velocity through the inlet valve Equation 2.15, and the width of the gap (assuming a valve seat angle of 45°), taking conditions at idle, $\overline{V}_p = 2$ m/s and $\rho = 0.4$ bar, we get a Reynolds number of about 6×10^3, which will still guarantee that the jet is turbulent (since the Reynolds number based on the turbulent velocity and length scales u and ℓ, which is the one that counts, will be in the neighborhood of $6 \times 10^2 \gg 1$, [96]). The Reynolds number will be much larger at higher piston speeds. This jet is issuing into an enclosed space, so it strikes the piston crown and the cylinder walls. It forms secondary vortices from the interaction with these surfaces, and these ultimately are dissipated or break up into turbulence. The impinging jet (as it is called) also forms turbulent boundary layers on the surfaces. The jet is itself unstable, and flutters back and forth irregularly, transforming the mean flow energy in the jet into turbulent energy.

We can make a crude estimate of the total energy introduced by the jet. If this energy were all to be converted without loss to turbulence, and spread uniformly through the cylinder, it would give us a rough estimate of the peak turbulence intensity we could expect during the inlet stroke. It would be a considerable overestimate, because we are assuming no losses; in Section 5.4 we saw that the losses (dissipation ϵ) are large. We can begin from our estimate for the mean velocity through the inlet valve, Equation (2.15).

$$\overline{V}_v = \overline{V}_p \left(\frac{b}{D}\right)^2 \frac{1}{C_i} \tag{5.14}$$

Now, we want to calculate the total kinetic energy that enters the cylinder in the jet through the inlet valve. Let us make our estimate when the jet has half-filled the cylinder, in the middle of the inlet stroke, when the turbulence from the jet is at its peak.

$$\text{Total kinetic energy} = \frac{\text{kinetic energy}}{\text{volume}} \text{volume} \tag{5.15}$$

The first term is

$$\frac{\text{kinetic energy}}{\text{volume}} = \rho_i \frac{\overline{V}_v^2}{2} \tag{5.16}$$

The second term is the total volume that has entered the cylinder, which is just half the displacement volume $V_d/2$. Hence, the total kinetic energy is

$$\text{total kinetic energy} = \rho_i \frac{\overline{V}_v^2}{2} \frac{V_d}{2} \tag{5.17}$$

If this energy is spread uniformly over the cylinder volume, we get a total energy of

$$\text{total turb energy in cylinder} = \rho_i \frac{u^2}{2} V_d \tag{5.18}$$

Equating expressions 5.17 and 5.18, we have

$$u = \frac{\overline{V}_v}{\sqrt{2}} \tag{5.19}$$

so that

$$\frac{u}{\overline{V}_p} = \left(\frac{b}{D}\right)^2 \frac{1}{C_i} \frac{1}{\sqrt{2}} \tag{5.20}$$

In [47] the peak value of $u/\overline{V}_p$ is given as about 10. Using Equation 5.20, estimating $C_i = 0.35$, and using the valve and cylinder diameters in [47], we obtain 12.4.

Obtaining a value that close from my simple calculation must be regarded as largely an accident. I neglected a great deal, notably the dissipation, which we have already seen in Section 5.4 is quite large, and should result in a considerable decrease in the turbulent velocity.

We have been considering an engine with one intake valve per cylinder, for the sake of simplicity. In a modern engine we are likely to have two. Roughly speaking, this will cut the velocity through the jet by a factor of two, and hence also cut the turbulent velocity by a factor of two. As piston speed drops, the turbulent velocity may no longer be large enough to provide a short enough burn time to avoid knock. The solution is to open only one of the intake valves at lower speeds, to bring the jet velocity, and the turbulent velocity up again. This is what Honda does, for example, in the VTEC engine (although that is also done to induce swirl).

Before we move on, we should consider decay of turbulence. The turbulent energy balance looks like this:

$$\frac{d\overline{q^2}/2}{dt} = P + T - D \tag{5.21}$$

where $\overline{q^2}/2$ is the turbulent mean kinetic energy per unit mass, P is production (the rate at which energy is being extracted from the mean flow and fed into the turbulence), T is transport (the rate at which energy is being moved to another place by the turbulence itself) and D is dissipation, $D = \epsilon$, the rate at which the turbulent kinetic energy is being converted to heat by viscosity. We do not need to concern ourselves with the mathematical forms of these terms now. During the inlet stroke, the turbulence away from the direct path of the inlet jet is receiving energy from both P and T. After the inlet valve has closed, the turbulence in the cylinder is relatively homogeneous, due to the effective transport properties of the intense turbulence during the inlet stroke. In fact, during the last third of the inlet stroke, the jet through the inlet valve is much weaker, and the production and transport are much less important. We have seen that the dissipation term can be very large. This term will result in a rapid decrease of the turbulent kinetic energy, in the absence of terms feeding the energy. We can make a rough estimate of how fast this can happen, if we write $\overline{q^2}/2 \approx 3u^2/2$, so that

$$\frac{d3u^2/2}{dt} = -\frac{u^3}{\ell} \tag{5.22}$$

We are supposing that the turbulence in all directions is equally intense (a property called isotropy). This is probably not true, but is close enough for a crude estimate – we are interested only in orders of magnitude here. We may easily solve Equation 5.22 if we assume that the turbulent length scale ℓ remains constant, to give

$$\frac{u}{u_0} = \frac{1}{1 + \frac{u_0 t}{3\ell}}$$

(5.23)

where u_0 is the initial value of u. If we take $u_0 = 10\overline{V}_p$, and $\ell = b/6 = S/6$, then we have

$$\frac{u}{u_0} = \frac{1}{1 + 20\frac{\overline{V}_p t}{S}}$$

(5.24)

If we ask, "How long does it take to reduce u/u_0 to one-half its initial value?," we easily find $\overline{V}_p t/S = 0.05$. If $\overline{V}_p t/S = 1.0$, the intensity u/u_0 is reduced to approximately 0.05 of its initial value. This is something like the correct value. Actually, the turbulent intensity is observed to drop to about 0.15 by the end of the intake stroke, and then only drops to 0.1 during the entire compression stroke. We have left out some important mechanisms: the attenuation due to the expansion during the last third of the intake stroke, and the amplification for the same reason during the compression stroke. ESP does this calculation correctly; all we are trying to do here is get a feeling for the order of magnitude for some of these effects.

5.6

INDUCING SWIRL AND TUMBLE

By the way in which the valves and ports are arranged, and the schedule of valve opening, mean flows can be induced in the cylinder. The flow into the cylinder is turbulent and the mean velocity is often smaller than the turbulent velocities. Motions like this are often called coherent, meaning that they are organization buried in the disorganized turbulence. When measurements are made, velocity patterns must be measured during a number of cycles (perhaps 20–30) and the results averaged to find this organized part. I reproduce here two figures from [105], Figures 5.3 and 5.4, which illustrate schematically tumble and swirl, two types of coherent motion that can be induced in the cylinder. It is also possible to induce a combination of the two motions. In fact, it is essentially impossible to generate swirl without inducing some tumble, so that the two are always associated. It is possible to

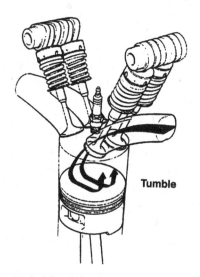

Tumble

Figure 5.3. Schematic of tumble from [105]. Reprinted with permission from 930821 © 1993 Society of Automotive Engineers, Inc.

generate tumble without swirl. However, tumble is always associated with other secondary motions, since it is generated by flow through two valves. We picture in Figure 5.5 schematically this secondary motion. This secondary motion has the effect of isolating the fluid entering through the $y > 0$ valve (in Figure 5.5) from the fluid entering through the $y < 0$ valve, something we will return to later. As we have indicated in the figure, the size of this secondary motion is approximately $b/2$, and the intensity is also about one-half of the intensity of the main tumbling motion. Swirl and tumble, or a combination of the two, represent the most general motion that can be induced at the scale of the cylinder. It is clear from

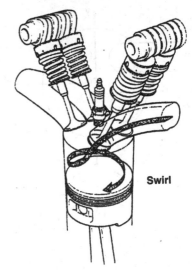

Swirl

Figure 5.4. Schematic of swirl from [105]. Reprinted with permission from 930821 © 1993 Society of Automotive Engineers, Inc.

the existence of the half-scale motion associated with tumble, that many more complicated motions are possible at smaller scale.

There are a number of reasons for inducing swirl and/or tumble. High turbulence levels at ignition produce higher effective flame speeds, and more reliable combustion at very lean air:fuel ratios, or with EGR. At normal air:fuel ratios, the higher speed will allow the flame front to reach the end gas before the chemical reaction resulting in auto-ignition has time to take place, permitting higher compression ratios without knock. This was the motivation in the early days of engine development, when fuel octane numbers were low. Now the motivation is more likely to be reliable combustion with very lean fuel:air ratios, or with EGR. In any event, one possible reason for swirl or tumble is to promote high turbulence levels at ignition.

Tumble can also be used for stratification. We will discuss this later.

As we have seen, the turbulence resulting from the conversion of the energy in the inlet jet decays rather fast, and not much is left at ignition. The idea of swirl or tumble is to encapsulate some of the momentum of the inlet valve jet in the organized motion (swirl or tumble), which is less dissipative than the turbulence (because of larger spacial scale), and hence will retain its energy longer. In addition, the vorticity in the tumble can be augmented by the compression process, or by squish in the case of swirl. Then, just before ignition the swirl or tumble can be induced to break up into turbulence, producing a much higher level of turbulence at ignition than would be present just from the decayed turbulence from the inlet jet.

Tumble appears always to break down to turbulence, because as the piston approaches TC, there is not room between the piston crown and the cylinder head for a vortex with a diameter of the order of b; only motions with scales of the order of the clearance height can survive, so the vortex breaks up into turbulence of this

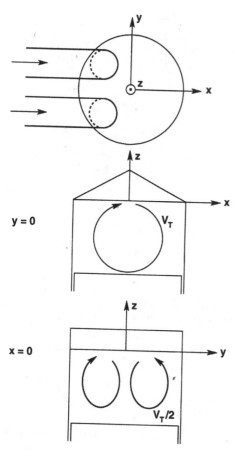

Figure 5.5. Schematic of secondary motion induced in tumble. Courtesy D. Haworth.

scale. In the case of swirl, if the combustion chamber is pancake-shaped, the swirl can survive through the burn, and we will talk about that in a moment. However, if the combustion chamber is a penta-head, with squish providing the transition from cylinder to head, and/or with the piston crown protruding into the head, the swirl must accommodate itself to the changing shape of the space available to it (from circular to rectangular, and increasingly narrow), and will also break up into turbulence.

In fact, in practical open-chamber four-valve pent-roof engines, squish is a minor factor. There is simply no room with four valves and a spark-plug to have any significant squish area. The same is true in modern two-valve engines. Hence, swirl survives to TC, and does not break up into turbulence. It is now clear from computational fluid dynamics and flow visualization that even in engines that were designed for high swirl, is principally the associated, unavoidable, tumble that yields the turbulence just before TC. There were apparent correlations between swirl and burn rate, but these appear to have held only because there was always tumble with the swirl.

The only exception to this is the high compression bowl-in-piston engine, where swirl plus the squish do result in spin-up, and the generation of turbulence.

The breakup of the tumble just before TC is not well understood, and is probably quite interesting from a fluid mechanical point of view [29], [38], [101]. Experiments on rotating flow in an ellipsoid show that rotation around the intermediate axis is not stable, and will change over to rotation around the minor axis if the ellipsoid is sufficiently flattened – that is, if the ratio of the axes is large enough. When the flow changes over, a strongly chaotic, turbulent, flow is generated. In an engine, this suggests that tumble, as the piston rises and the space available becomes more flattened, will change over to swirl, with the generation of strong turbulence. The losses will be large, probably enough so that the swirl generated will be very small.

The other possible reason for inducing swirl is to stratify the charge; that is, to keep the fuel-rich charge segregated, so that it does not mix with the remainder of the air in the cylinder. In order to do this, it is necessary to damp some of the turbulence, selectively. It turns out (as we shall see) that one of the effects of swirl is to damp strongly the turbulence near the edge of the swirl, and if the combustion chamber is pancake-shaped, so that the swirl can persist until ignition, the swirl vortex at ignition consists of a high turbulence core surrounded by a very low-turbulence annulus. If the fuel-air cloud can be induced to enter this low-turbulence annulus (and it can, as we will see) then it will stay there, and will not be spread throughout the cylinder. Ignition will not be a problem, because the fuel/air mixture is locally near stoichiometric in the cloud. At ignition, the flame front will at first be in the low-turbulence annulus, and will spread slowly, but it will soon leave for the leaner mixture in the high-turbulence core, where it will burn reliably and spread rapidly.

Another type of stratification associated with swirl is important in direct injection engines. The hot residual gases are considerably lighter than the cold incoming charge (perhaps 0.25 of the density), and will move toward the axis of rotation. Fuel vapor is of considerably higher density than air, and with direct injection engines it is possible to obtain a rich mixture with a density ratio of perhaps 1.5–2.0, which will gravitate to the outside of the swirl. Whether these will stay put depends on whether the density gradients are large enough to damp the turbulence, which would otherwise mix them. These density gradients are definitely large enough.

Even if the swirl is broken up by squish and a penta head, if this can be delayed until close enough to ignition, a similar scenario will hold. At ignition, the turbulence will be high, so flame propagation will be rapid. What is important is to not provide enough time for the turbulence to spread the fuel cloud before ignition. If breakup can be delayed until roughly 15 CAD before ignition, then the resulting turbulence cannot spread the fuel cloud more than 1/3 of the available vertical distance (perhaps 0.25 S) before ignition.

Swirl and tumble are usually present together. The combination of tumble and swirl results in a tilted axis of rotation of the secondary motion in the cylinder, and this tilted axis precesses. So long as the tilt of the axis is not too great (say, 1:3), the suppression effect on the turbulence near the edge of the swirl is still present [80].

Both swirl and tumble are normally specified by a swirl ratio, or tumble ratio. In either case, the angular velocity of the solid-body rotation with the same angular momentum as the actual velocity distribution in the swirl or tumble is compared to the angular velocity of the crankshaft.

$$R_s = \frac{\omega_s}{2\pi N}$$

$$R_t = \frac{\omega_t}{2\pi N}$$

(5.25)

where R_s and R_t are respectively the swirl ratio and the tumble ratio, and ω_s and ω_t are respectively the angular velocities of the solid-body rotations that have the same angular momentum as the real swirl or tumble flow. N has units of revolutions/sec.

Production engines usually have values of the swirl or tumble ratios between 1.0 and 2.0. Experimental engines achieve numbers up to perhaps 6.0. If we call the tangential velocity at the edge of the solid-body rotation v_θ, so that $\omega_s = 2v_\theta/b$, then we can write

$$R_s = \frac{\omega_s}{2\pi N} = \frac{v_\theta}{b/2} \frac{S}{2\pi NS} = \frac{2}{\pi} \frac{S}{b} \frac{v_\theta}{\overline{V_p}} \tag{5.26}$$

Hence

$$\frac{v_\theta}{\overline{V_p}} = \frac{\pi}{2} \frac{b}{S} R_s \tag{5.27}$$

For a square engine, roughly 3/2 of the swirl ratio R_s gives the ratio of the tangential velocity to the mean piston speed. Thus, in production engines we expect tangential velocities below three times the mean piston speed, while experimental engines may achieve nine times the mean piston speed. We can say the same things for tumble.

The swirl and tumble ratios are normally measured in steady flow on a test rig consisting of the cylinder head and valves, with a tube replacing the cylinder. A paddle wheel, or similar device is placed at the exit from the tube, and the rotation of the device is used as a measure of the swirl or tumble. This approach is sound for swirl, but is less satisfactory for tumble. It is used because measurement of the actual swirl or tumble in a motoring engine, or an operating engine, is extremely difficult and expensive. This can be done, using various optical techniques (particle image velocimetry [80], laser doppler velocimetry [100]) if an experimental engine has been specially prepared for optical access to the cylinder. The match between the steady flow swirl ratios and the swirl ratios measured in a motoring engine is good if the actual velocity distribution is integrated to obtain the angular momentum. For example, [80] had a steady flow swirl ratio of 6.0, and the PIV results build from near zero at TC on the intake stroke to about 6.5 at BC, then fall slowly to about 5.0 at TC on the compression stroke. On the other hand, estimating the swirl ratio from the maximum tangential velocity gives low values; this gives 2.93 instead of 5.0 for the conditions of [80]. In [100], at TC the swirl ratio (estimated from the maximum tangential velocity, measured by LDV) was 1.27, while the steady flow value was about 2.6. At 150 CAD before TC, the estimated value was a little closer, at 1.75. It appears that estimating the swirl ratio in this way (from the maximum tangential velocity) gives about 0.6 of the true value. This number would vary for different velocity distributions in the swirl; however, the velocity distributions are usually quite similar, at least for open chamber four-valve pent-roof engines. Presumably the same remarks can be made about the tumble ratios. For two-valve engines, and/or bowl-in-piston designs, it is more difficult to generalize.

Swirl and tumble ratios can be obtained quite accurately for both production and research engines using computational fluid dynamics. This is not a cheap solution, however, because it requires the construction of a grid for the inlet manifolds, ports, valves and combustion chamber, unless this has been constructed for some other purpose.

Although the swirl velocity field is often more complex during the intake stroke and early on the compression stroke, by TC of the compression stroke, the mean velocity field has settled down to a single vortex, centered in the cylinder. I

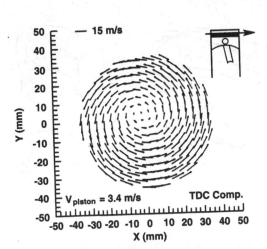

Figure 5.6. Measured velocity distribution in pancake-shaped combustion chamber at TC, $b = 92$ mm, $S = 86$ mm, at $\overline{V}_p = 3.4$ m/s. The inlet valve has a 120° shroud. From [80]. Reprinted with permission from 952381 © 1995 Society of Automotive Engineers, Inc.

include here (Figure 5.6) PIV measurements of the velocity field at TC from [80]. Bear in mind that this is a particularly simple engine.

5.6.1 Lift Strategies

As can be seen from Figure 5.3, the four-valve head lends itself to producing tumble, which can be optimized by modifying the angle of the inlet runners. This is probably one explanation for the popularity of the four valve head, even though other valve arrangements might have heat transfer or Mach index advantages.

On the other hand, if both inlet valves are opened, it does not produce swirl (presuming that the head is reflectionally symmetric about the plane between the two valves). In order to produce swirl, it is necessary to keep one of the inlet valves closed, or nearly so. In [105], they experimented with variable valve lift in a 79 mm × 76.2 mm engine at $\overline{V}_p = 5$ m/s. With equal lift they attained $R_s = 0$; with one valve closed, they obtained $R_s = 0.85$. The $R_t = 0.52$ with the valves equally open, but rose to about $R_t = 1.0$ with one closed. In the Honda VTEC engine, for example, this strategy is used to obtain high swirl and tumble. Usually, the valve that is not opening is actually opened just a crack to allow the fuel sprayed by the injector into the port to enter the cylinder. Because the air flow is very small, this is sometimes used to keep a fuel-rich cloud near the cylinder head, to produce a stratified charge.

5.6.2 Port and Valve Configurations

Swirl can also be generated by the configuration of the inlet port, or by shrouding the valve, or masking it. I reproduce here Fig. 5.7 from [47], showing various inlet port configurations, which produce similar values of swirl ratios, between

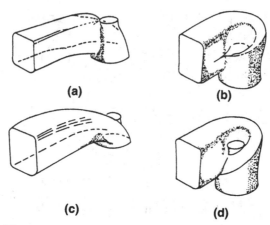

(a)   (b)

(c)   (d)

Figure 5.7. Different types of swirl-generating inlet ports: (a) deflector wall; (b) directed; (c) shallow-ramp helical; (d) steep-ramp helical [97]. Reproduced by kind permission of the American Society of Mechanical Engineers.

2.5 and 2.9. The deflector wall port uses the port inner side wall to force the flow preferentially through the outer periphery of the valve opening, in a tangential direction. The directed port brings the flow toward the valve opening in the desired tangential direction (see [47]). These various ports pay various penalties in lower discharge coefficients. The helical ramp ports appear to have higher discharge coefficients, because the whole periphery of the valve open area can be used [47]. The directed port, with its straight passage, has the most restricted flow area, and the lowest discharge coefficient. Since only one wall is used in the deflector wall port, the area is less restricted, and the discharge coefficient is higher.

Port design could probably be approached rationally. For example, it seems possible that there is a design producing a minimum discharge coefficient for given swirl or tumble, a design that could be found by variation of design parameters using computational fluid mechanics. However, in real engine design the final (suboptimal) port choice is made based on non-fluid-mechanical considerations of packaging, fuel injector targeting, and the like, so that concern with optimal ports is probably not productive.

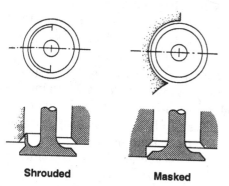

Shrouded   Masked

Figure 5.8. Shrouded inlet valve and masked cylinder head approaches for producing net in-cylinder swirl angular momentum [47]. Copyright © 1988 by McGraw-Hill, Inc. Reproduced with permission of The McGraw-Hill Companies.

Another way of inducing swirl is by shrouding or masking the valve. Figure 5.8 from [47] illustrates shrouding and masking.

Shrouding is often used on experimental engines, but never on production engines, since it requires that the valve be prevented from turning. Normally, a valve is turned slightly by the lifter each time it is raised, and this has advantages in uniformizing valve head temperature and seat wear. Some special provision has to be made to prevent this rotation if shrouding is used. Masking avoids this problem, and can easily be included in a production engine. However, both shrouding and masking have the disadvantage that the effective valve open area is reduced, which reduces the volumetric efficiency and increases the Mach index.

5.7
EFFECT OF COMPRESSION

To understand what happens to either swirl or tumble during compression, we have to talk about fluid mechanics for a moment. Conservation of angular momentum is the principle that we will apply. We are all familiar with the illustration of the ice skater who spins on one toe with her arms outstretched, and then pulls in the arms, reducing the radius of gyration (as it is called), and increasing the angular velocity substantially. This is conservation of angular momentum – her angular momentum with arms out and with arms in is the same, but the moment of inertia is larger with the arms out, and smaller with the arms in, so the angular velocity is smaller with the arms out, and larger with the arms in.

In the engine cylinder, the swirl vortex is being squashed lengthwise, while the tumble vortex is not having its length changed (we will consider the effect of squish separately in a moment – for now, assume there is no squish). However, the situation in the engine cylinder is complicated by the fact that the density is changing, as well as the length of the vortex. We need a more general statement of conservation of angular momentum with density change. Deriving this here would take us far from our subject; the interested reader can consult a good fluid mechanics text, such as [12]. What we need is called Cauchy's equation, which states (in a simplified form) that, in an inviscid fluid, if a vortex of initial vorticity ω_0, initial density ρ_0 and initial length ℓ_0 is stretched to a length ℓ_f, and changed to a density ρ_f, then the final value of the vorticity ω_f is given by

$$\frac{\rho_0}{\rho_f} \frac{\omega_f}{\omega_0} = \frac{\ell_f}{\ell_0} \tag{5.28}$$

The same equation can also be obtained by starting with the integral form of angular momentum conservation, and assuming solid-body rotation. Equation 5.28 can now be applied in the engine cylinder.

5.7.1 Effect on Swirl and Tumble

First, consider swirl. If we start at the beginning of the compression stroke, ℓ_0 is the stroke plus the clearance height, $S + h$; at TC, ℓ_f is simply h. The ratio

$$\frac{\ell_f}{\ell_0} = \frac{1}{r} \tag{5.29}$$

where r is the compression ratio. The density ratio, on the other hand is

$$\frac{\rho_f}{\rho_0} = r \tag{5.30}$$

so that the ratio of the final to the initial vorticity is

$$\frac{\omega_f}{\omega_0} = r \cdot \frac{1}{r} = 1 \tag{5.31}$$

Hence, there is no change in the value of the swirl vorticity during compression.

Tumble, on the other hand, is another story. Here, the axis of the vortex is transverse to the cylinder, so that there is no change in the length of the vortex during the compression (again, assume no squish). Thus,

$$\frac{\omega_f}{\omega_0} = \frac{\rho_f}{\rho_0} = r \tag{5.32}$$

so that the effect of the density change is to spin up the vortex quite strongly. The tumble vortex will break up before the piston reaches TC. Also, in both cases, the vortices are losing energy to the cylinder walls during the stroke, and we have not taken this effect into consideration. Nevertheless, it is clear that the tumble vortex should contribute a considerably higher turbulence level at ignition, if this is what is desired.

We can make a crude estimate of the effect of friction on a swirl vortex. The wall shear stress is approximately

$$\tau_w = \rho u_\tau^2 = \rho \frac{u_\tau^2}{v_\theta^2} v_\theta^2 \approx \rho \frac{v_\theta^2}{900} \tag{5.33}$$

where u_τ is the friction velocity (defined by the first part of Equation 5.33), and we have made the crudest possible approximation: $u_\tau / v_\theta = 1/30$, which is approximately valid over a very wide range of Reynolds numbers (see [96]). We are, after all, only trying to find out if this effect is important or not. Using the fact that the rate of change of the angular momentum is equal to the applied torque, and supposing that the piston is stationary (another crude assumption which will greatly simplify the problem), we can obtain the equation

$$\frac{dv_\theta}{dt} = -\frac{v_\theta^2 8}{900b} \tag{5.34}$$

This may be immediately integrated to give

$$\frac{v_\theta}{v_\theta^0} = \frac{1}{1 + \frac{8tv_\theta^0}{900b}} \tag{5.35}$$

where v_θ^0 is the initial value of v_θ. If we take $S = b$, and assume that $v_\theta^0 = 3\overline{V}_p$ (corresponding to $R_s \approx 2$), a reasonable value, we obtain

$$\frac{t\overline{V}_p}{S} = 37.5 \tag{5.36}$$

as the time required for the initial tangential velocity to drop by one-half. That is, it will take some 38 strokes for this to happen. Despite all the approximations, it is clear that this is not an important effect. There were more approximations that we did not mention – we neglected the torque on the ends of the mass of gas. However, it is obvious that adding these will not make enough difference to make this important.

Tumble is more difficult to treat, because the losses are not so simple to parameterize. We have a cylindrical vortex with its axis transverse to the cylinder, and as the piston comes up, and the vortex is compressed between the piston crown and the cylinder head, the geometry is difficult to model in any simple way. The only thing that is clear is that the losses are considerably greater than they are for swirl. This is clear, for example, in [80], where the initial swirl ratio is 6, and the tumble ratio is 2. At the end of the compression stroke, the swirl ratio should still be 6, but the tumble ratio should have gone up to 16. However, from the observed tilt of the vortex axis, the tumble ratio is evidently still of order 2. The simplest way to view this, is that the losses are so great that all the additional energy (put in by the compression) is converted into turbulence.

If there is squish we can take that into account using the same equation. Consider the effect on swirl. We can do the calculation in two steps. First, consider a simple compression without squish. The combustion chamber is cylindrical and the same diameter as the cylinder. The piston will stop with a relatively large clearance volume (a low compression ratio) to leave room for the squish. According to our equation (above), Equation 5.31, there will be no change in the vorticity. Now bring in the sides of the combustion chamber to produce the squish. During this phase there will be no change in the length of the vortex. If, for example, the radius is reduced to one-half its value, the area is reduced to one-quarter of its initial value, so the density increases by a factor of four, and hence the vorticity is up by a factor of four. I have sketched this in Figure 5.9. We can apply the compression and the squish sequentially, because Equation 5.28 relates only the initial and final states. Recall that squish is negligible except in bowl-in-piston or bowl-in-head engines.

When tumble and swirl are present simultaneously, Equations 5.30 and 5.31 are simultaneously valid.

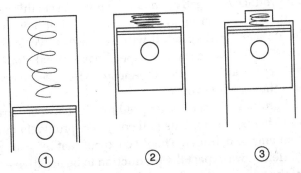

Figure 5.9. Cartoon of a swirl vortex being first compressed and then squished.

5.7.2 Effect on Turbulence

A swirling flow has interesting effects on turbulence. The radial distribution of angular momentum can act dynamically like a stably or unstably stratified temperature distribution in a gravitational field (see [96]).

We will have to stop for a moment and talk a little about turbulence dynamics. See Equation 5.21 and the discussion immediately following it. We need to concern ourselves only with the production. In a mean flow on circular streamlines, with tangential velocity U, the production term is

$$P = -\overline{uv}\, r \frac{\partial}{\partial r}\left(\frac{U}{r}\right)$$

(5.37)

where $\overline{uv}$ is called the Reynolds stress. It is the mean value of the product of the fluctuating turbulent velocities in the streamwise and radial directions. This product is not zero, because the velocities are correlated – that is, when one goes up, the other tends to go up (or perhaps down) also. We will get back to that in a ' moment.

The term $r\frac{\partial}{\partial r}\left(\frac{U}{r}\right)$ corresponds to the velocity gradient in a parallel flow. If the mean flow is a solid body rotation, so that the speed U is proportional to the radius, then this vanishes. There is no shear in a solid body rotation.

In a turbulent flow, lumps of fluid are continually changing position. Although they are interacting with their surroundings and interchanging momentum with the fluid with which they are in contact, they are trying to conserve angular momentum also, just as a lump of fluid in a parallel flow is trying to conserve linear momentum, while interacting with its surroundings. We can make a very crude estimate of $\overline{uv}$ if we suppose that a lump having the local tangential velocity of the swirling flow is displaced radially outward (a positive fluctuation v) to a new, larger radius, and approximately conserves its angular momentum during this displacement. Conserving angular momentum would mean that its velocity would be $\propto r^{-1}$, so its swirl speed will drop as it moves to its new radius. A velocity field in which $U \propto r^{-1}$ is called a free vortex; this has the same angular momentum at all values of r. When the displaced lump arrives at its new (larger) radius, if the lump finds its swirl speed to be slower than the mean swirl speed of the surroundings at the new radius , it produces a negative *velocity fluctuation u*; this gives a value of the product $\overline{uv} < 0$. On the other hand, if the displaced lump finds itself moving faster than the local mean swirl speed, it will have a positive fluctuation $u > 0$. Hence, we expect $\overline{uv} < 0$ if the mean swirl speed has a radial slope less negative than that of a free vortex, and $\overline{uv} > 0$ if the mean swirl speed has a radial slope more negative than that of a free vortex.

Consider radial profiles of mean swirl speed that lie between a free vortex and solid body rotation (that is, $U \propto r$). This will cover any profile that we will find in a swirl vortex in an engine cylinder. Then we expect that $\overline{uv} < 0$, and the mean shear $r\frac{\partial}{\partial r}\left(\frac{U}{r}\right) < 0$. Hence, we expect the production to be negative – that is, it will take energy out of the turbulence and put it back into the mean flow. It will act to suppress the turbulence. It will be a stabilizing influence, like a stable temperature distribution.

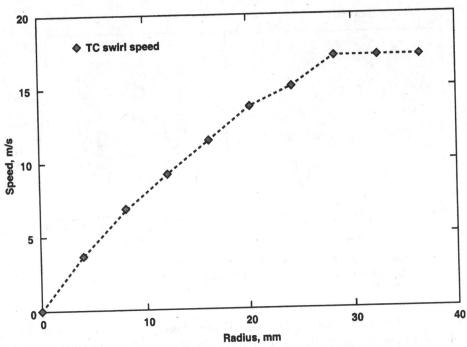

Figure 5.10. Distribution of tangential speed relative to the vortex center in the swirl of Figure 5.6.

In Figure 5.10 I have plotted the distribution of tangential speed in the vortex of Figure 5.6.

It is clear that the swirl vortex is almost in solid body rotation out to a radius of approximately 28 mm, and that the tangential velocity has an essentially constant value after that. The solid body rotation is due to a high turbulence level. Intense turbulence is like a very viscous fluid – it will actively transport momentum in such a way as to try to bring the mean profile to solid body rotation (a state of zero shear). We expect that there will be very little production out to a radius of 28 mm, and strong negative production beyond that point.

We can make a crude estimate of the strength of this negative production. We have determined that $\overline{uv} < 0$; u and v should be well correlated, if the fluctuations in u are essentially caused by the fluctuations in v combined with the conservation of angular momentum. Hence, we should expect $\overline{uv} \approx -u^2$, where u is the r.m.s. value of the turbulent velocity fluctuation. Let us take $U \propto r$ out to a certain radius, say near 28 mm and constant $= U_0$ thereafter. Then the production will be equal to (in the region where it is negative and non-zero, for $r > 28$ mm),

$$P = -\overline{uv}\, r \frac{\partial}{\partial r}\left(\frac{U}{r}\right) \approx -u^2 \frac{U_0}{r} \tag{5.38}$$

We should compare this with the dissipation, which is given by $D = \epsilon = u^3/\ell$: let us take $\ell = b/6$ and evaluate the ratio P/D at $r = b/2$:

$$P/D \approx -\frac{u^2 U_0 \ell}{r u^3} = -\frac{U_0}{u}\frac{1}{3} \tag{5.39}$$

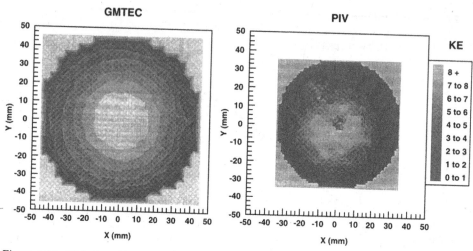

Figure 5.11. Turbulent kinetic energy – calculated and measured – at TC compression midplane corresponding to Figure 5.6 [80]. Reprinted with permission from 952381 © 1995 Society of Automotive Engineers, Inc.

Looking at Figure 5.10, we see that $U_0 \approx 17$ m/s. Figure 5.11, which we will describe in a moment, gives the peak value of the measured turbulent fluctuating velocity in this flow at $u = 3.46$ m/s. As a result, we have $-U_0/3u \approx 17/10.4 \approx 1.6$. This estimate is extremely crude, but it suggests that the suppressive effect of this negative production term may be important. In this region we have, according to our estimate, destructive forces (negative production plus dissipation) $= -2.6\epsilon$. This means (approximately) that the fluctuating velocity in the peripheral region will decay to a value 1/2.6 of the fluctuating velocity in the core in the same time, or that the fluctuating kinetic energy in the peripheral region will be about 15% of the core kinetic energy after the same decay period. This is numerically approximately what Figure 5.11 shows.

Let us look at Figure 5.11, reproduced from [80], which presents the measured and calculated turbulence distributions corresponding to Figure 5.6. Look first at the measured distribution. In the center of the cylinder, the turbulence level is roughly 3.4 m/s, or about the mean piston speed. The turbulence level is thus about twice what can be achieved without swirl and tumble. It is also evident that, from a radius of about 26 mm out to the cylinder wall (at least to the resolution of the measurements – a word about that in a moment) there is very little turbulence, in the neighborhood of 1 m²/s². This is exactly the region in which we have estimated that there is a negative production roughly 1.6 times the dissipation. Note: if there is no turbulence, there is no negative production. The negative production we have calculated was present when there was turbulence, and it is what killed the turbulence, but now that the turbulence is dead, the negative production has died also. In Figure 5.11 we are looking at the flow just before TC, but the value of negative production we have calculated corresponds to the conditions during the compression stroke.

The calculated distribution does not reproduce this. The turbulence falls off in a sort of Gaussian shape close to the cylinder wall, but does not fall to zero.

This is because the turbulence model used cannot reproduce the behavior of the turbulence in a swirling flow. Developing a turbulence model that is capable of reproducing such behavior is on the frontiers of research at the moment (see, for example, [55, 83, 84]; these references refer to rotating flows, but rotating and swirling flows are closely related).

Very close to the wall there is a turbulent boundary layer. This layer is not resolved by the PIV technique. The cylinder has a bore of 92 mm, which gives a radius of 46 mm. The PIV data are limited to a circle of radius roughly 35 mm, as can be seen in Figure 5.11, so there is a layer next to the wall of thickness roughly 11 mm that is not observed. In this region, there is a relatively thin turbulent boundary layer trying to propagate against the stabilizing negative production at its edge. This boundary layer is unstable, and is dominated by shear, unlike the flow just above it, which is dominated by rotation. Boundary layers like this in the atmosphere, propagating upward against buoyant stably stratified fluid, are well-understood, but our boundary layer in this rotating flow is not, and we cannot predict the thickness of this boundary layer, other than to say that it will be thinner than a similar boundary layer in a non-rotating flow. The turbulence in the boundary layer will be relatively weak, approximately $v_\theta/30$, which makes it about one-tenth of the intensity of the turbulence in the center of the cylinder. It will also be of relatively small scale, roughly 1/3 the thickness of the boundary layer, which is a small fraction of the bore. The effective diffusivity in the boundary layer, $u\ell$, will therefore be roughly 2% (or much less) of the diffusivity in the center of the bore.

In this flow of [80] the initial tumble ratio was 2, while the swirl ratio was 6. As we have seen, the swirl ratio after compression should still be 6, but the tumble ratio should be 16 ($r = 8$). However, as we noted above, the tumble ratio stays roughly constant, indicating that all the additional work done on the tumble vortex by the compression has been transformed to turbulence. However, it is evident from Figure 5.11 that the negative production in the outer 1/3 of the radius has managed to kill the turbulence put in by the breaking up tumble as well.

In several recent works [9], [44], [57], detailed budgets of angular momentum, mean and turbulent kinetic energy have been calculated in the engine cylinder. These suffer from being modeled turbulence, and fairly simple models. However, this would not be a problem – our analytical models are also quite simplistic. Unfortunately, the budgets presented are global budgets, integrated over the cylinder volume, so that no information is available on local differences in production.

It is clear that there is a lot going on here that would repay closer attention.

5.8
CHARGE STRATIFICATION

We have already hinted that this turbulence suppression associated with swirl can be used for charge stratification. In support of our conclusions so far, in [15] they conclude that "...swirl tends to reduce mixing along the cylinder axis," and "...radial mixing is slow and axial mixing is dominated by compression due to the ascending piston." These authors find that the primary cause of charge stratification is the fuel droplets that enter through the inlet valves and impinge on the cylinder wall opposite the valve (having been sprayed into the port by the injector). They will then evaporate from the cylinder wall. Due to the turbulence suppression by the swirl, this cloud of fuel vapor mixes only slowly with the rest of the charge in the cylinder, and remains in the top of the cylinder. In order for this to happen, it is necessary to have a certain amount of tumble in the entering flow, so that the flow is directed against the cylinder wall, and not straight down onto the piston crown. This tumble will break up as the piston approaches TC; say, between 620 and 660 CAD (where TC of combustion is 0 CAD), and ignition is at 680 CAD, so there may not be time for the additional turbulence generated by the breakup of the tumble to disperse the fuel-rich cloud which has been segregated up to now by the turbulence suppression of the swirl, if the turbulence generated by the tumble is not too strong. However, [100] note: "A low tumble ratio is a necessary condition for obtaining axial mixture stratification in the cylinder. In other words, the smaller the tumble ratio is, the more clear axial mixture stratification is observed in the cylinder."

There is a primary cloud of fuel droplets and vapor that does not impinge on the cylinder wall, and is carried down into the cylinder. This mixes quickly with the cylinder contents and is much leaner than the rich cloud resulting from evaporation from the wall.

All this makes clear that direct injection would be highly desirable, since it would permit introduction of the fuel cloud close to TC, when there would not be sufficient time to spread it uniformly.

One powerful stratification mechanism involves the migration of regions of differing density under the influence of centrifugal acceleration. A stoichiometric mixture of isooctane C_8H_{18} is about 18% denser than air. We can evaluate the centrifugal acceleration using Equation 5.27 to determine v_θ. If we take $R_s = 2$, $b = S$ and $\overline{V}_p = 5$ m/s, we find that $a_c \approx 6$ km/s^2, a considerable figure. However, the question is, how far can it move a fuel cloud in the available time? The acceleration felt by the fuel cloud is given by

$$a_c \frac{\Delta\rho}{\rho} \tag{5.40}$$

If we take $\Delta\rho/\rho = 0.2$, and $R_s = 2$, we find that it takes about one-third of a stroke for the fuel cloud to move half of the cylinder radius under the influence of the centrifugal acceleration. This suggests that the centrifugal forces can be quite effective in segregating the fuel cloud in the cylinder, and will tend to hold the cloud of fuel evaporated from the cylinder wall near the wall (see Problem 5).

We must not forget residual or recirculated exhaust gases also. These will be much hotter than the incoming gases, and hence of much lower density. Even if the exhaust gases have lost half the temperature difference from the incoming gases, they will still have a temperature of 800 K, giving a density about 0.375 of the incoming gases. As a result, $\Delta\rho/\rho \approx 0.6$, reducing the time to move half the radius to about 1/5 of a stroke. Consequently, the exhaust gases will be expected to move smartly to the center of the cylinder.

5.9
SQUISH

Squish, as we have seen, can intensify swirl toward TC, which is surely helpful. Also, as the piston approaches TC, the charge trapped between the piston crown and the squish area (the bumping clearance [7]) is forced out in a jet into the center of the cylinder. This is a turbulent jet. This turbulence occurs very near TC, and may be too late and too weak to do much transport of the charge before ignition. It will probably also not have a great influence on the burn, since it does not spread very far [7]. In addition, the intensification of the swirl due to the squish occurs too close to TC to have much influence on turbulence suppression, since the negative production due to the swirl needs time to act [7].

With the exception of bowl-in-piston or bowl-in-head designs, as used for direct injection engines, the squish effect tends to be small in practical engine designs. Pent-roof chambers and near flat pistons do not break down swirl effectively, because the amount of squish possible is minimal.

5.10
POLLUTION

The automobile is not the only source of atmospheric pollution; in production of oxides of nitrogen it is about on a par (47%) with other sources; in production of hydrocarbons it is a little below other sources, at 39%, while in production of carbon monoxide it is the major source at 66%; in production of carbon dioxide, it is responsible for something like 14% of fossil fuel burning worldwide [41]. The automobile has several features that make it essential to control. Automobiles are everywhere, and are close to the ground, so they constitute a distributed, low-level source. Particularly in urban areas, this distributed pollution is difficult for the atmosphere to remove. In addition, where there are cars, there are people, so that the pollution is associated with high population density, the worst possible situation. If this is combined with an unfortunate atmospheric and geographic situation that tends to confine the pollution, and prevent its being washed away by the weather (as in Los Angeles and Denver), we have a recipe for disaster. Politically, since there is a vehicle for every one or two adults in the United States or western Europe [103], a politician risks a negative response from a large fraction of the electorate with attempts to place controls on cars and car use. Although car drivers do not constitute an organized group that donates large amounts of money to campaigns, automobile industries do, so it is politically somewhat dangerous to be too zealous in pursuing this source.

Automobiles are responsible for several different types of pollution. We will deal with several of these below. The principal pollutant from the automobile is not noxious and has no short-term effects, but may have extremely serious long-term effects; that is carbon dioxide. Anything fueled by a carbon-based compound will produce carbon dioxide as one of its products, and carbon dioxide is a greenhouse gas – that is, its presence in the atmosphere lets in the short-wave radiation, but prevents the long-wave radiation from leaving. There is no question that the levels of atmospheric carbon dioxide are increasing steadily, and that this is due to human activities. It seems incontestable that this will do something to the climate. Changes in rainfall patterns, possible el-Niño-type events, triggering of a new ice age [23], and global temperature increase are some of the possibilities. The global temperature seems to have been increasing, but not as smoothly as the CO_2, for unexplained reasons. It fact, it seems possible that 78% of the observed temperature increase over the last 100 years can be explained by changes in sunspot count and associated changes in brightness [75]. There are a number of unresolved issues regarding the question of just what, and how much, the increase of CO_2 will do to the climate. For example, will a small increase in temperature cause an increase in rainfall and cloud cover, increasing the planetary albedo, with a resultant fall in solar radiation absorbed? This is just one mechanism that is not included in any of the models, and which may be very important, but which we cannot quantify at the moment.

If the temperature increases significantly, the oceans will expand, and the water level may rise. How much, depends on whether the temperature rise is confined to the surface layers of the ocean, or not, and to the effect on precipitation. For example, it is not clear whether the polar ice caps will melt – [103]. If the ice caps melt, they will add to the rise. On the other hand, if the changes in atmospheric moisture cause increased accumulation of ice at the poles, the sea level might not rise. If there is a rise, the consequences would be serious, resulting in flooding of coastal areas like New York City, and of small island nations.

Each kilogram of fuel produces some three kilograms of CO_2. This works out to about 20 kg of CO_2 per 100 km for an average car. To reduce the CO_2 production substantially, the fuel consumption must be reduced. Other fuels have different numbers of carbon atoms, and hence produce different amounts of CO_2, but the differences are not great. For example, methane is one of the most attractive alternatives, because it has just one carbon atom, as opposed to the eight of isooctane. To produce the same amount of energy, methane produces only 0.89 the mass of CO_2 produced by gasoline.

Although the consequences are not completely clear, the possibilities are sufficiently frightening that there is a general feeling that anthropogenic CO_2 emissions should be reduced or at least controlled. Toward the end of 1997 an international meeting was held in Kyoto to attempt to reach agreement on a reduction schedule. The automobile is only a minor player; all sources will have to be controlled, and the negotiation to determine how much each source will be controlled is political and delicate. Some start is being made, but more than a billion people want the benefits that development and energy consumption provide, the automobile among them, and they are not taking kindly to being told (by the developed world) that they can't have them.

There are several things that can be done to alleviate CO_2 production by the automobile. Fuel consumption could be substantially cut with engines of the current type if the vehicle were much lighter. The majority of fuel consumption is spent in acceleration (and then dissipated to heat when the vehicle is braked). A much lighter vehicle could run with a much smaller engine and still have interesting performance. Drag coefficients could be cut approximately in half, at most [50, 76], although a factor of 2/3 is probably more realistic; this would reduce fuel consumption at higher speeds. High fuel economy prototype vehicles, which are mostly quite light, and some of which have low drag coefficients, have fuel consumptions in the neighborhood of 3–4 L/100 km [76]. The New Car Corporate Average Fuel Economy of the domestic fleet is currently about 27.5 miles per gallon, corresponding to a consumption of about 8.6 L/100 km, or more than double that of the high fuel economy prototypes [4]. Space-age plastics and aluminum are beginning to be used in automobile construction and can help to reduce total vehicle weights considerably.

As an example, a consortium of suppliers in the plastics, adhesives, plastic molding and tooling disciplines, led by the Chrysler Liberty group, has recently designed the Composite Concept Vehicle, an ultra-light-weight vehicle designed for sale in China and in the Third World [68]. It has a thermoplastic body with the color molded in, a ladder frame, high ground clearance, a 19 kW air cooled 2-cylinder V engine, a mass of 545 kg, has a fuel consumption in the neighborhood of 4.5 L/100 km, and is designed to sell for US$6,000 (see Figure 5.12). Getting the price down this low is quite a *tour de force*, but I suspect that, considering the average annual income in China (US$620 in 1995 [6]), it is probably still not low enough. At 4.5 L/100 km, its output of CO_2 per km is about half that of the

Figure 5.12. Composite Concept Vehicle, and the Citroën 2CV (Deux Chevaux) that inspired many of its features. Photograph © by J. Frenak [68].

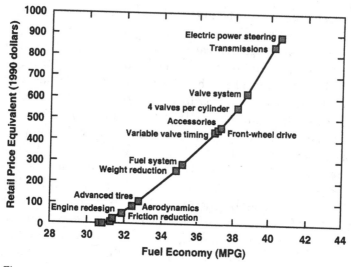

Figure 5.13. Fuel economy-technology cost curve [76].

average member of the U.S. domestic fleet; however, the addition of one such vehicle for every few people in China would double the global number of motor vehicle registrations. Even if we were all driving cars like this, the total production of CO_2 by cars would not then be reduced [103].

More or less conventional techniques (many of which I have referred to already in this book) can make a big difference: engine redesign, friction reduction, aerodynamics, advanced tires, weight reduction, variable valve timing, more efficient accessory drives, better valving and valve actuation and many other minor improvements, have the potential to make about a 30% improvement in fuel consumption [76]. See Figure 5.13.

Everything has its cost, and the cost in 1990 dollars (US) of these various modifications is estimated as roughly US$900 per car [76]. It is not clear that the buyer is prepared to pay that now.

The hybrid electric vehicle has a relatively small combustion engine driving an alternator; this engine runs continuously, at its speed of maximum efficiency, and charges batteries. The car is driven by electric motors, using the batteries. In high pollution regions, the engine can be shut off. Such a vehicle has a higher overall efficiency and hence lower fuel consumption and CO_2 production (both about one-half of the current U.S. domestic fleet, [102]), since the engine is operating at peak efficiency at all times, and braking is regenerative, so that the kinetic energy is not wasted. Overall efficiency is limited by the energy-conversion efficiencies (thermal-mechanical-electrical and back again), which multiply. It is also limited by the performance of its batteries, which are expensive, heavy and do not last many cycles of charging/discharging.

The same comments can be made about the all-electric vehicle, which must carry even more heavy, expensive, short-lived batteries, and has the added inconvenience of a short range and a long charging time, and at present an infrastructure that does not provide frequent convenient charging stations. I can personally speak for 50 years of news stories excitedly describing breakthroughs in battery technology, which would make the electric car a practical reality, but which never came to anything. There is even a conspiracy theory that claims that the American automobile industry is responsible for quietly crushing this nascent battery technology. I do not believe that. I think battery development is difficult given the resources we are prepared to commit to it, and the batteries we have are the best we can manage at the present time and maybe for a long time to come.

Recently [102], some progress has been made in the U.S. with a process that takes gasoline and converts it to hydrogen and carbon dioxide at room temperature, called a reformer [102]; the carbon dioxide need not be released to the atmosphere (although it is not clear what the developers propose to do with it), and the hydrogen is then combined in a fuel cell with oxygen to produce electricity and make water vapor, again at room temperature. This has the potential to use the infrastructure which is already in place to deliver gasoline to drivers, can be designed to release to the atmosphere only water vapor, and produces only 10% of the oxides of nitrogen (because of the low temperature of the reaction) which we shall see are serious contaminants. In addition, the efficiency is about double that of the IC engine, so that the fuel consumption is about halved. This is because the energy available from the reaction $2H_2 + O_2 \rightarrow 2H_2O$ is substantially greater than that available from the usual reaction. Hence, even if the CO_2 were released to the atmosphere, the quantity produced would be halved. On the down side, it is currently very expensive, and at least a decade of development time is forseen in optimistic news reports, which probably means (realistically) considerably more. It is hard to envision a billion cars in China and Africa and South America fueled by this expensive, high-tech process, with the supporting technological infrastructure in place, at least not soon. If the CO_2 is not released to the atmosphere this process has considerable promise, since it then produces no greenhouse gases or oxides of nitrogen. Note that the transient response of many types of fuel cell is poor, and fuel cell vehicles may also require substantial battery packs.

Recently, at the 1997 Frankfurt Motor Show, Mercedes-Benz unveiled a very similar reformer-fuel cell combination, but methanol based [27]. It requires no battery pack and no hydrogen storage, and the fuel cell and reformer are small enough to fit under the floor and in the rear of the A-class vehicle. Mercedes-Benz is suggesting commercialization in six years, when they hope to sell 100,000 of these vehicles, beginning with municipal fleets. Price was not mentioned.

People like the automobile as we have known it. As I have suggested in the Preface, many of us are emotionally entangled with the familiar automobile. In addition, huge industries have an enormous investment in our kind of automobile. All this means that if change comes it will come slowly and we will all be dragged backward, kicking and screaming, into the future. Nevertheless, there are many pressures for change or possibly phase-out of the internal combustion engine, and CO_2 emissions are only one of them. Four-six billion people each having a personal vehicle will surely be a problem, no matter what other changes are made. The various other emissions will have to be brought under control. Analysts at General Motors believe that by 2010 an accurate assessment of global climate change should be available, and that petroleum prices may start escalating, as petroleum demand approaches and exceeds capacity; by 2050, significant global climate change should have occurred, and petroleum production capacity should be dead [34]. There are other sources of fuel (shale, coal, biomass, etc). In any event, conditions will probably need to get substantially worse before there is the financial incentive for any action, which suggests that the internal combustion engine as we know it will be with us for some decades, but with increasing pressure to reduce emissions and fuel consumption.

5.10.1 Atmospheric Chemistry

The internal combustion engine produces several other varieties of pollution which have more serious short-term consequences, in that they are responsible for urban air quality deterioration.

Carbon monoxide is extremely toxic. Levels of 0.01 percent by volume in air cause headaches and loss of mental acuity; levels of 0.1 percent cause death. It is a serious problem in urban areas; in forty-four major U.S. metropolitan areas with a combined population of some 30 million people, the national carbon monoxide air quality standard is not being met [41]. In 1988, approximately 67% of the CO emissions in the U.S. came from transportation [41]. In addition, CO contributes to the elevation of atmospheric levels of ozone and methane, and hence contributes to the formation of photochemical smog [41].

The emission of oxides of nitrogen and of hydrocarbons contributes to the formation of photochemical smog, as in the Los Angeles basin and the Denver basin, Mexico City, Beijing, Santiago and increasingly many urban areas. In Bangkok, motorists spend 44 days a year idling in standstill traffic; at least a million people in Bangkok (out of a 10 million population) received treatment for smog-related respiratory problems in 1990 [103].

The formation of photochemical smog is a slow, complex process, and I cannot go into it in detail here. I will give a simplistic sketch: NO from the engine undergoes a photochemical reaction with the hydrocarbons (such as ethylene and butane) produced by the engine to produce NO_2. [*Photochemical* means that the input of energy in the form of solar radiation is necessary to make the reaction go.] Then the NO_2 undergoes a further photochemical reaction to produce NO and O. Then the O combines with O_2 to form O_3 [103]. O_3 is ozone, the principal constituent of photochemical smog [41]. Ozone causes eye irritation, coughing and chest discomfort, headaches, upper respiratory illness, increased asthma attacks and reduced pulmonary function; it reduces crop productivity, and kills Ponderosa and Jeffrey pines and eastern white pines [41].

5.10.2 Chemistry in the Cylinder

Air is three-quarters nitrogen. At room temperature, nitrogen is quite inert, and does not participate in many reactions. However, at the temperatures of combustion in the engine cylinder, nitrogen becomes quite reactive. In particular, both N_2

and O_2 molecules dissociate, and participate in two chain reactions (known as the Zel'dovich mechanism) which form NO. There are also other reactions involving the hydrocarbon fuel (for example, that produce OH molecules which combine with nitrogen to form NO) which contribute to the NO production [103].

In Figure 5.14 I show the concentrations in the exhaust of a spark ignition engine of NO, CO and unburned hydrocarbons. The NO production is primarily a function of temperature. The adiabatic flame temperature peaks for mixtures slightly richer than stoichiometric, and falls for both fuel-rich and fuel-lean, even for gaseous hydrocarbon fuels.

In addition, NO is not stable at room temperature, and if the reaction rates remained the same as they are at combustion temperatures the NO produced would not last long. However, as the temperature drops during the expansion stroke, and when the exhaust valve opens, the reaction is *frozen* – that is,

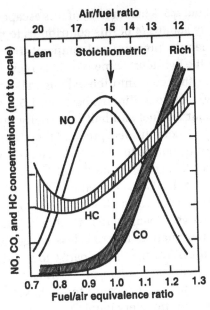

Figure 5.14. The HC (unburned hydrocarbons), CO and NO concentrations in the exhaust of a spark ignition IC engine as a function of the equivalence ratio Φ. Φ is defined as the actual fuel:air ratio divided by the stoichiometric fuel: air ratio. If it is greater than 1, the mixture is rich (excess fuel); if it is less than 1, it is lean (excess air) [47]. Copyright © 1988 by McGraw-Hill, Inc. Reproduced with permission of The McGraw-Hill Companies.

the reaction rate drops to a very small value. Recall (from the section on thermodynamics in Chapter 1) that the reaction rate contains the Arrhenius factor $e^{-E/RT}$, where E is the activation energy (a constant), R is the universal gas constant, and T the absolute temperature. As the gas temperature falls, the value of the Arrhenius factor falls very rapidly. Although the compounds are not stable at room temperature, the time for the reaction to take place may be measured in weeks.

Unburned hydrocarbons are produced in the cylinder primarily from the quenched regions between the piston and the cylinder wall, above the top piston ring, as well as inside the spark plug, around the valves and in corners; that is, regions that are surrounded by too much chilled surface area, so that their temperature is reduced below the point of ignition. In addition, the charge is poorly mixed, so that there are regions of mixture that are too rich to burn. When the mixture is lean, the excess air manages to burn some of these regions, so that the concentration of unburned hydrocarbons in the exhaust drops. The situation is worse when the mixture is rich; the quenched layers thicken, and the incombustable rich regions are larger, so that the amount unburned in the exhaust increases. According to [25], roughly 9% of the fuel entering the cylinder escapes normal combustion.

About 4.5% (about half) of this escaped fuel is oxidised later in the expansion, but only about 1/4 of this contributes to the *imep* (because the oxidation takes place when the pressure and temperature have fallen). The other 4.5% is not oxidised in the cylinder. Some of this is recycled (as residual gases or as blowby). Of the part that goes into the exhaust, about one-third is oxidised in the exhaust port and manifold. Finally, between 1.5% and 2% (say, 1.8%) remains. Of this, half is unburned fuel and half is partially reacted fuel components. We are simplifying here a complicated process; there are many routes which can lead from the incoming charge to the exhaust, and it requires a detailed flow chart to make them clear. This can be found in [25].

Finally, CO is a product of incomplete combustion. In an abundance of oxygen, CO would combine with the O_2 to form CO_2. Hence, when there is an excess of oxygen (lean) there is relatively little CO produced. As the mixture gets richer, so that there is less and less oxygen available, the amount of CO produced increases. Again, this has partly to do with the poor mixing of the charge. The same effect would happen in a perfectly mixed reactor. However, here it is exaggerated somewhat because the richest regions in the poorly mixed charge produce disproportionately more CO.

Note that even under very lean operating conditions, the engine still produces NO, CO and unburned hydrocarbons.

5.11

LEAN BURN

The upper and lower air:fuel mass ratios for ignition of gasoline in air are approximately 42.8 and 3.2 [99]. However, ignition of a mixture substantially leaner than stoichiometric is not reliable. The ignition limit means that a mixture leaner than 42.8 is essentially impossible to ignite, but an engine will run only very unevenly, with many misses, at mixtures well beyond stoichiometric toward the lean limit. For example, Figure 5.15 from [58] shows the torque fluctuation level as a function of air:fuel ratio. This figure refers to a homogeneous charge lean-burn engine, and is rather qualitative. Torque fluctuation is a measure of misfiring. The lean combustion band, in which NO_x is down and torque fluctuation has not yet risen, can be moved by optimization of swirl and tumble, squish area, stratification of the charge, a high energy ignition system, fast combustion (presumably dependent on turbulence level) and decrease of fuel droplet size [58].

Note that lower engine-out NO_x does not necessarily result in lower vehicle emissions for a lean-burn engine. After-treatment efficiency by the catalyst drops rapidly with excess air. So far as I am aware, no lean-burn system can meet the 50-state U.S. NO_x standards.

Thermal efficiency is improved in lean burn for several reasons: the exhaust gases have a composition much closer to air, and a lower temperature, and both of these increase the specific heat; the induction of excess air reduces the pumping losses; heat loss is reduced because the exhaust gas temperature is lower [58].

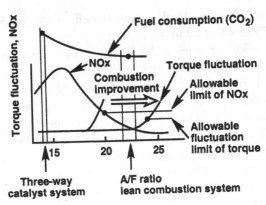

Figure 5.15. Effect of air:fuel ratio on NO_x emission, fuel consumption and torque fluctuation [58]. Reprinted with permission from 940307 © 1994 Society of Automotive Engineers, Inc.

We have already seen that squish probably will not do much of itself to improve the situation, but it can serve to break up swirl to produce high turbulence levels. Our options are basically use of swirl and/or tumble to increase turbulence at TC, or swirl (and tumble) to produce charge stratification (see Section 5.8).

Injection of the fuel directly into the cylinder (Direct Injection, or the Gasoline Direct Injection–GDI–engine) can produce a highly stratified charge. This is often combined with a bowl-in-piston or bowl-in-head configuration, with strong squish. Here squish can play a direct role in containing the fuel and augmenting the turbulence. Now we must be concerned with spray-flow interaction, fuel vaporization, spray-wall flow interaction, mean flow advection of the fuel cloud, thermodynamic conditions, and much more. We will deal separately with the direct injection engines later. Here we will confine our attention to the slightly stratified, port-fuel-injected (PFI) engine.

For charge stratification, we can examine for example [15]. The overall air:fuel ratio is in the neighborhood of 19–20 in this experimental engine. This engine has a four valve penta head. In lean operation only one inlet valve is significantly opened (the other is just cracked to drain the fuel sprayed into the port). There is mild squish to break up the swirl at TC. Swirl and tumble ratios are in the neighborhood of 2. At ignition (680 CAD) a cloud of stoichiometric mixture extended to 5 mm below the spark plug. The stoichiometric cloud ignites readily; once ignited, even a lean mixture will usually continue to burn, and the high turbulence level helps to assure this and to keep the burn time down. A lean mixture has a considerably longer burn time, other things being equal [47].

Although the laminar flame speed S_L drops substantially, this is not primarily responsible for the longer burn time. The effective flame speed is primarily controlled by the turbulent velocity. To see this, consider: in an engine with $r = 8.5$, $\overline{V}_p = 3.98$ m/s and $\Phi = 1$, the laminar flame speed is $S_L = 0.75$ m/s, while the turbulent velocity is $u = 2$ m/s [47], giving an effective speed of propagation of the brush of $u + S_L = 2.75$ m/s. A reduction in the laminar speed to, say, 0.23 m/s under very lean conditions [47] will reduce the effective speed of propagation of the brush from 2.75 m/s to 2.23 m/s.

The dominant effect producing the longer burn is that lean burn results in lower product temperature, which results in higher product density. In combustion, the flame front propagation speed is the sum of two speeds: the turbulent effective flame speed relative to stationary gas, and the gas velocity, which is produced by the generation of high temperature product by the reaction. Conservation of the mass flowing through the flame front (in coordinates set in the flame front) says that the gas velocity and the effective flame speed must be related by the density ratio on the two sides of the flame. High density out means low speed gas flow out, and *vice versa*. Consequently, if the product density is higher (due to the lower temperature) the gas velocity is lower. Lower gas velocity means a lower speed of expansion of the flame.

5.11.1 Honda VTEC-E 1.5 L SOHC 16 Valve Four-in-Line

The Honda VTEC-E is designed to run lean only during light load acceleration and cruising below 2,500 rpm, at an air:fuel ratio of 24:1. During hard acceleration, and at engine speeds of more than 2,500 rpm, the engine runs at stoichiometric conditions, and uses a three-way catalyst.

The variable valve timing and lift arrangements used in this engine have already been described in Figure 2.23 and associated text. Under lean operation, one intake valve is just cracked, while the other is opened wide, to generate strong swirl. This engine has a penta head, with about 13% squish. The charge is stratified.

In lean operation, the engine has an EPA rating of 48 mpg (4.9 L/100 km) city, 55 mpg (4.28 L/100 km) highway. For operation in California to meet that state's more stringent NO_x standards, it operates stoichiometric, and its EPA ratings are 44/51 (5.35/4.62). Hence, the lean burn is worth about a 10% improvement in fuel economy [67].

5.11.2 Toyota Carina 4A-ELU 1.6 L DOHC 16 Valve Four-in-Line

This engine is marketed only in Japan. The basic approach is exactly the same as that used by Honda in the VTEC-E engine, but instead of the complex Honda valve gear, Toyota uses a swirl-control flapper valve that shuts off the flow of air and fuel to one of the two intake runners below 2,800 rpm [67].

5.11.3 Mitsubishi Mirage 4G15MPI-MVV 1.5 L SOHC 12 Valve Four-in-Line

This Mitsubishi Vertical Vortex (MVV) engine is also available only in Japan. This is how Mitsubishi refers to tumble. Each cylinder has two intakes and one exhaust, with the spark plug located where the second exhaust valve would ordinarily be. *Car and Driver* [67] says: "The fuel is stratified by injecting it into only the intake runner that is pointed at the spark plug. The…swirls…generate strong turbulence…." The engine (see Figure 5.16) appears to have a penta head with squish. As we have seen, this configuration should generate strong turbulence as the tumble is broken up as the piston approaches TC. The interesting thing is the claim that the charge is stratified. That only seems possible if the turbulence in the outer third of the tumble vortex is suppressed during compression, just as it is in the swirl vortex.[1] I have not been able to find measurements in the literature of the turbulence in this configuration, but the fuel cloud would not remain opposite the spark plug until ignition unless the turbulence were suppressed. From a fluid-mechanical point of view I would expect both swirl and tumble vortices to behave in the same way. The tumble vortex will have a somewhat more difficult time, because of the awkward shape of the cylinder, which should result in a somewhat thicker boundary layer near the cylinder walls. The breakup of the tumble vortex near TC would provide

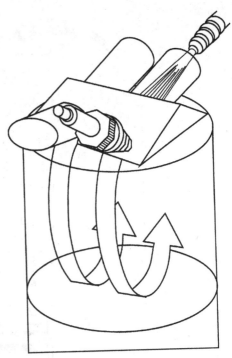

Figure 5.16. A sketch of the Mitsubishi Mirage Vertical Vortex engine [67]. Reproduced by permission, Mitsubishi Motor Sales of America, Inc.

a high turbulence level which would assure reliable and fast combustion. This engine delivers 20% better fuel economy on the bench than the standard engine.

5.11.4 Mazda Surround Combustion 2.0 L DOHC 16 Valve Four-in-Line

This engine has four spark plugs. It operates on a single central plug at idle, on three circumferentially placed plugs during cruise, and on all four plugs during heavy acceleration. The engine appears to have a penta head. It has a domed piston that appears to provide some squish toward the periphery (see Figure 5.17) [67]. There appears to be no swirl, but there is probably tumble associated with the penta head. This would be broken up by the domed piston for a high turbulence level at ignition. No attempt appears to have been made for charge stratification. Rather, with multiple spark plugs, the likelihood that the mixture will be ignited in considerably increased; suppose that the probability of ignition with one plug is 0.65, with four plugs it is 0.99 that at least one will fire. In addition, the high turbulence level promotes a short burn. If all plugs were firing, the short distance the flame front must travel would also make for a short burn, but we cannot count on that – only one may be firing, and then the distance the flame front must travel is completely normal.

---

[1] But see Figure 5.5 for another possible contributor.

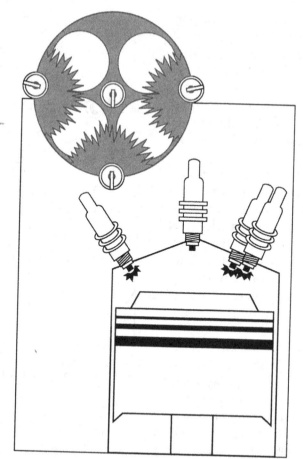

Figure 5.17. Sketch of the Mazda surround combustion engine [67]. Reproduced by permission; Mazda Motors.

5.12

GASOLINE DIRECT-INJECTION ENGINES

The direct-injection spark-ignition engine injects the fuel directly into the cylinder, where it is ignited by a spark plug. These engines have compression ratios in the normal range for a spark-ignition engine. Such engines have a number of advantages (as well as some disadvantages) all of which we will discuss below. Briefly, they offer high resistance to knock, as well as really stratified charge and various thermodynamic advantages, which combine to give substantial improvement to *bsfc*.

The direct-injection spark-ignition engine is not new. The first such engines appear to have been developed by Knut J. E. Hesselman, a Swedish engineer, with many Swedish, other European and U.S. patents for IC engines, fuel injection devices, etc., from 1925 to 1938. He patented a number of configurations, with bowl-in-piston combustion chambers, pre-combustion chambers of various sorts, and various devices for inducing swirl. His patents contain drawings

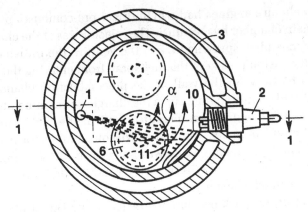

Figure 5.18. Sketch of combustion chamber with swirl and GDI, from [46].

(see Figure 5.18) showing swirl (note the shrouded valve, 11 on the figure) carrying the cloud of fuel spray around to the spark plug. Note the deflector 10, attached to the top of the piston, to prevent the raw fuel from landing on the cylinder wall. This should be compared with the modern Figure 5.19, showing current mixture preparation strategies, which are very similar.

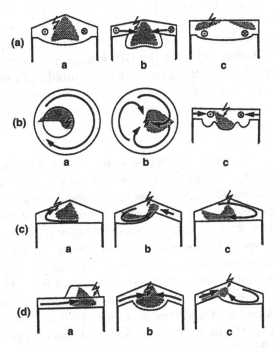

Figure 5.19. Mixture preparation strategies for GDI combustion systems [32]: (a) swirl-based systems with centrally mounted injector; (b) swirl-based systems with centrally mounted spark plug; (c) tumble-based systems; (d) squish-based systems. Reprinted with permission from 970627 © 1997 Society of Automotive Engineers, Inc.

One of Hesselman's engines had a cylindrical pre-combustion chamber connected to the main chamber by a venturi. The injector was at the closed end of the pre-combustion chamber, spraying toward the venturi; the inrush of air through the venturi as the piston rose kept the fuel spray from entering the cylinder. The spark plug was halfway up the wall of the precombustion chamber. This was therefore a stratified charge engine, since the charge was confined to the precombustion chamber. This engine was manufactured in the U.S. under license by the Waukesha Motor Company, Waukesha, Wisconson, a major manufacturer of diesel engines. The Hesselman engine was often used for farm tractors, since it was economical (for reasons we will see below) and would run on a wide variety of fuels, due to its knock resistance.

The direct-injection spark-ignition engine (Gasoline Direct Injection, or GDI, engine) is attractive because it makes possible much greater charge stratification than is possible in a port-fuel injection, or PFI engine. This can be brought about by the timing of the charge injection, which can be quite late in the compression stroke, and by the design of the injector, which (in combination with the fluid mechanics of the flow in the cylinder) can effectively isolate the fuel spray. Since much greater charge stratification is possible, much greater fuel economy is possible. The attractiveness of the GDI engine has meant that interest in such engines is recurrent. In just the last eleven years (the only ones for which I have information) three major GDI engine programs have swept through the industry: a 4-stroke program in the late 1970s – mid 1980s; a 2-stroke program in the late 1980s – early 1990s; and the current 4-stroke program of the mid- to late 1990s.

As we have indicated, although the GDI engine has great advantages with regard to charge stratification, there are continuing problems with pollution. Unless EGR is used, when these engines are running very lean (under partial load) the effectiveness of the catalytic converter is very much compromised. None of these engines can meet the 50-state U.S. pollution requirements at the present time.

GDI engines have a number of advantages: they need not be throttled (although most existing prototypes still are), which would reduce pumping losses; in addition, in an unthrottled stratified charge engine the same amount of energy is released into a larger mass of gas, producing a lower temperature rise, and lower heat losses. The injection of the fuel into the cylinder, and its evaporation, results in cooling of the charge. This increases the volumetric efficiency, and also permits lower octane requirements or higher compression ratios. (The charge cooling is relative to PFI engines, where evaporation takes heat from the port and valve walls; in carburetted engines, much of the heat for fuel evaporation also comes from the carburettor and manifold structure.) All of this results in up to 30% improvement in BSFC [107]. In addition, at low load, the charge is prepared just before ignition, so that the time available for the auto-ignition reaction is very short; this also permits lower octane requirements.

Under low loads, the GDI engine operates by injecting a small fuel charge late in the compression stroke, and by management of the flow in the cylinder and the design of the spray nozzle, contrives to keep this fuel charge separated from most of the air in the cylinder. This often involves a complex combustion chamber design, involving squish and a bowl in the piston crown or in the cylinder head. As the load increases, the injection timing is moved earlier and earlier into the intake stroke, and the fuel mixes more and more with the air in the cylinder, until at full load the charge is homogeneous. The fact that the fuel cloud is occupying a changing volume and location in the cylinder as the load changes, while the ignition location (the spark plug) is fixed, presents a problem. The fuel spray and the air motion must be managed to keep a cloud of ignitable vapor near the spark plug no matter what the load. In some designs the change from late injection to early injection is not gradual, and other solutions, such as double injection, must be used in the intermediate load range. These are described in [107]. Computer control of injection volume and timing is relatively straightforward; by contrast, mechanical control (in the days before computers) was considerably more difficult, and this difficulty was probably responsible for the Hesselman engine having limited application, typically in constant load situations.

GDI engines also have advantages in better transient response, requiring less acceleration enrichment, more rapid starting, with less cold enrichment required, reduced cold-start hydrocarbon emissions and reduced CO_2 emissions. However, hydrocarbon emissions are generally excessive [107]. Also, GDI engines tend to produce particulates, like Diesel engines, and for the same reasons: the fuel spray contains a range of droplet sizes. The largest droplets do not have time to evaporate, and for this reason are incompletely burned, leaving a carbon particle.

A conventional, homogeneous charge, lean-burn PFI engine produces decreasing levels of NO_x as the mixture becomes leaner, which occurs due to reduction in the reaction zone temperature. However, in a GDI stratified charge engine, the local temperature of the reaction zone remains high, since some areas are stoichiometric or slightly rich. NO_x production is high in these areas. As a result, the NO_x production of a GDI engine without EGR is similar to that of a PFI engine, even though the GDI engine may have an air:fuel ratio near 50 at low load.

Since a conventional three-way catalyst cannot be used for an engine such as a GDI that operates very lean, another technique must be used to remove NO_x. EGR is widely used to reduce in-cylinder NO_x production. It operates as a diluent, just like air, reducing combustion temperatures. However, it degrades the thermal efficiency somewhat, because of the presence of CO_2 and H_2O molecules, which have a higher specific heat, and produce a lower polytropic index. In a GDI engine, stable combustion is possible with a much larger EGR ratio than is possible in a PFI (homogeneous, or nearly so) engine, because the charge in the neighborhood of the spark plug is near stoichiometric. A PFI homogeneous engine is limited to something like 15% EGR, while a stratified charge GDI engine can operate somewhat above 30%.

Apart from careful management of the flow in the cylinder, this requires careful design of the injector. Droplet size and droplet velocity will determine droplet evaporation rate and droplet penetration [107]. That is, how far the droplets will travel into the air in the cylinder. In the PFI engine, a large part of the fuel charge is deposited on the walls of the port and the back of the inlet valve, where it must be evaporated. If the full advantage of the GDI engine is to be achieved, the fuel spray must not hit the cylinder wall to form a liquid fuel film there. GDI engines generally use a common rail fuel injection system – that is, a common high-pressure fuel manifold feeding the injectors, which are electronically controlled by the engine computer. The pressure in the manifold is roughly 5–7 MPa, a compromise – higher pressure produces smaller droplets, but greater penetration. Pressure during the starting cycle is usually much lower, in the neighborhood of 500 kPa. The typical injector produces a conical sheet of fuel, which is given a spin about its axis. The sheet breaks up into droplets very quickly. Mean droplet sizes in the neighborhood of 20 μm are achieved [107]. The advantage of the swirl nozzle is that it produces a relatively narrow range of droplet sizes (if the mean size is 20 μm, the maximum is something like 50 μm; a hole-type nozzle, on the other hand, has a maximum size in the neighborhood of 100 μm).

The management of the flow in the cylinder, together with the injector pattern, presents a number of possibilities. In Figure 5.19 [32] I show a number of possible

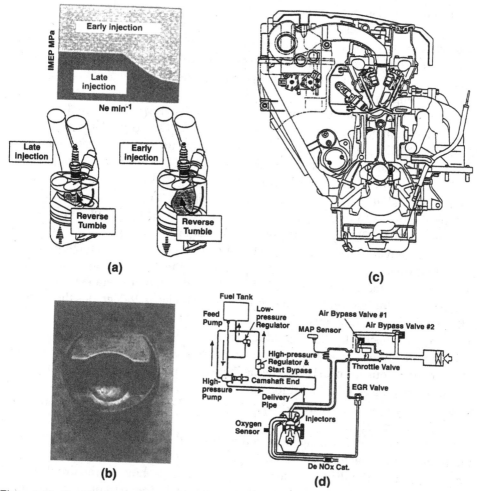

Figure 5.20. The Mitsubishi GDI combustion system [107]: (a) Basic concept of mixture preparation; (b) Piston photograph; (c) Engine cutaway; (d) System layout. Reprinted with permission from 970627 © 1997 Society of Automotive Engineers, Inc.

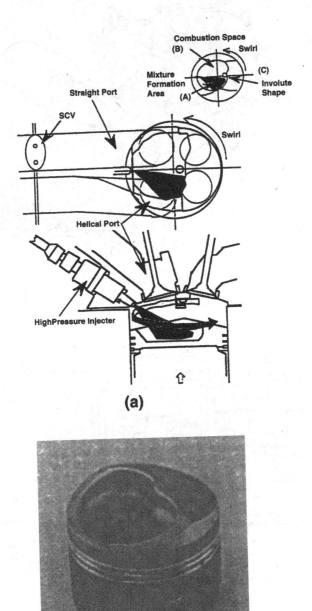

(a)

(b)

Figure 5.21. The Toyota GDI combustion system [107]: (a) Basic concept of mixture preparation; (b) piston photograph. Reprinted with permission from 970627 © 1997 Society of Automotive Engineers, Inc. (continued.)

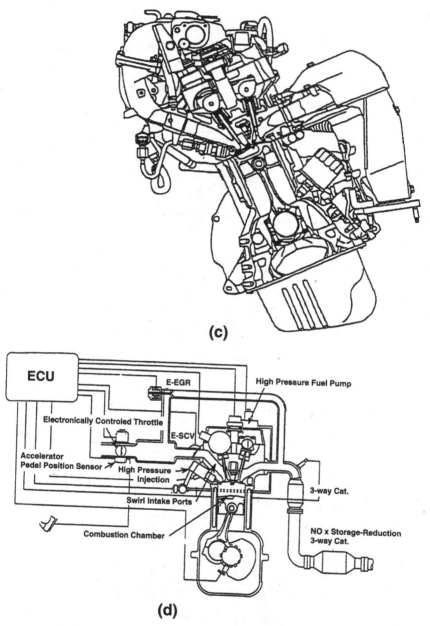

(c)

(d)

Figure 5.21 (continued). The Toyota GDI combustion system [107]: (c) Engine cutaway; (d) system layout. Reprinted with permission from 970627 © 1997 Society of Automotive Engineers, Inc.

patterns. Generally, ignition stability is maintained by positioning the spark plug on the periphery of the fuel spray.

In Sections 5.12.1 and 5.12.2 I will describe two GDI stratified charge engines, from [107]. This by no means exhausts the available engines, although the ones I will describe are the closest to production. Nissan, Isuzu, Ford, Ricardo,

Mercedes-Benz, Texaco, MAN, VW and several others have research – prototype engines under development. Information on these can be found in [107].

5.12.1 Mitsubishi GDI Engine

This is a four-cylinder, four-valve-per-cylinder, DOHC engine with a pentroof combustion chamber and a 12:1 compression ratio (see Fig. 5.20). It is based on a parent homogeneous charge PFI engine of lower compression ratio, but the GDI engine a different cylinder head and piston, and injectors. The piston has a spherical cavity, and the intake ports are angled to produce a tumble opposite to the usual direction. The fuel pressure is 5 MPa. The air:fuel ratio is up to 40. It uses 30% EGR, and a de-NO_x catalyst. The BSFC is approximately 35% better than the parent PFI engine. The 0–100 km/h acceleration time was improved by 5%, as was the volumetric efficiency. This, plus the increased compression ratio increased the power output by 10% [107].

5.12.2 Toyota GDI Engine

This is a four-cylinder, four-valve-per-cylinder, DOHC engine with a pentroof combustion chamber and a 10:1 compression ratio, also based on a parent PFI homogeneous engine (see Fig. 5.21). The piston has a heart-shaped cavity. One intake port is straight and one helical. An electronically controlled butterfly swirl control valve is located upstream of the straight port. When the swirl control valve closes off the straight port (for light load), and the flow is forced to enter through the helical port, the swirl ratio is 2.1. The helical port uses a variable valve timing intelligent cam-phasing system. For light load, the injection is late in the compression stroke. For heavy load the swirl control valve is opened, and the fuel is injected during the intake stroke, producing a homogeneous mixture [107]. The air:fuel ratio is up to 55. The engine uses electronically controlled EGR up to 40%. The BSFC was improved over the parent engine by about 30%, and the 0–100 km/h acceleration was improved by about 10%. The torque was improved about 10% in the low to mid-rpm range. At low engine loads, the intake valve is opened earlier, increasing the overlap. This results in internal EGR, increasing the total EGR ratio while requiring less from the external EGR system, and contributing to NO_x reduction.

5.13

PROBLEMS

1. To explore the moving average concept, consider a "mean flow" plus a "turbulence" given by

$$u(x) = A \sin \kappa_1 x + B \sin \kappa_2 x \qquad (5.41)$$

for $0 \leq x \leq 2\pi/\kappa_1$, where $\kappa_2 \geq \kappa_1$. The term multiplied by A is intended to be the "mean flow," and the term multiplied by B is intended to be the "turbulence." It is possible that $\kappa_2 \gg \kappa_1$, but not necessarily. Define an average velocity by

$$\bar{u}(x) = \frac{1}{L} \int_{-L/2}^{+L/2} u(x+\xi)\, d\xi \tag{5.42}$$

where $L \leq 2\pi/\kappa_1$. This is a moving average. To avoid problems at the ends of the interval, suppose that $u(x)$ repeats periodically at both ends of the interval. Then define a fluctuating velocity as

$$u(x) - \bar{u}(x) = u'(x) \tag{5.43}$$

- Now compute $\overline{u'^2}(x)$ for $\kappa_2 L/2 = 10\pi$, $\kappa_1 L/2 = \pi/10$, so that $\kappa_2/\kappa_1 = 10^2$;
- for $\kappa_2 L/2 = \sqrt{10}\pi$, $\kappa_1 L/2 = \pi/\sqrt{10}$, so that $\kappa_2/\kappa_1 = 10$;
- and for $\kappa_2 L/2 = 10^{1/4}\pi$, $\kappa_1 L/2 = \pi/10^{1/4}$, so that $\kappa_2/\kappa_1 = \sqrt{10}$.
- Discuss the results.

2. When the coherent motions (swirl and/or tumble) break up into turbulence at about 22.5 CAD before TC on the compression stroke, the turbulence immediately begins to decay. The decay can be used to infer the turbulent length scale ℓ, which appears in $\epsilon = u^3/\ell$. Consider a particular engine with a compression ratio $r = 8$, and suppose that the turbulent length scale is constant. In this engine, you observe initially (at 22.5 CAD before TC on the compression stroke) $u_0/\overline{V}_p = 5.0$, and at TC $u_f/\overline{V}_p = 0.633$. The turbulence production is small and probably negligible, and the flow is probably homogeneous enough to neglect the transport. Use Equation 5.22 and estimate the size of the turbulence length scale ℓ as a fraction of the clearance height h.

3. The velocity entering through the valve is given approximately by Equation 5.14. If the port is arranged so that this velocity is essentially tangential to the cylinder, it will produce swirl. Taking account of the drop in jet velocity as the jet spreads, we should be able to predict easily attainable values of the swirl ratio. The jet from the valve will spread as it travels away from the valve. The spreading rate is constant, and is always the same [96]. Call the effective radius of the jet λ, and the effective radius of the jet at the valve λ_0, where $A_f = \pi\lambda_0^2$, where A_f is given by Equation 2.2. Then

$$\lambda = \lambda_0 + 0.067x \tag{5.44}$$

where x is the distance traveled. Take the distance traveled as approximately one circumference, once around the cylinder πb. Take $A_f = 0.735\pi D^2/4$, (corresponding to the XK engine) and take $D/b = 0.44$ corresponding to Figure 2.7a. The jet velocity times the jet area ($\pi\lambda^2$) must be constant. Use Equation 5.14 for $\overline{V}_v$. Use Equation 5.27 to relate tangential velocity (the jet velocity after one trip around the cylinder) to the swirl ratio. Show that the swirl ratio attainable in this way is a little over 2.0. To do much better than this probably requires shrouding or masking.

4. Consider an engine run on methane, CH_4, with specific calorific value of 50 MJ/kg. Calculate how many kilograms of CO_2 are produced per megajoule of energy produced, and compare with the combustion of isooctane, C_8H_{18}, specific calorific value 44.6 MJ/kg.

5. If the incoming charge is density-inhomogeneous, the centripetal acceleration of the swirling flow in the cylinder will cause it to segregate, heavier
material at the outside and lighter in the middle. We can make a guess at how
long this will take. Consider a swirling flow with a swirl ratio of 2.0. Using
Equation 5.27 to relate the swirl velocity to the swirl ratio, estimate the swirl
velocity, supposing that the engine is square, $b = S$, and that the mean piston
speed is 5 m/s. Compute a centripetal acceleration, $2v_\theta^2/b$. Suppose we have
a density anomaly of order one (that is, a density between one-half and two
times the mean density). This will migrate at the centripetal acceleration.
See how long it will take to move half-a-radius, $b/4$, relative to the time for
a stroke, $S/\bar{V}_p$.

6. Consider a GDI injector, operating at a pressure of 6 MPa. The spray enters
the cylinder approximately at TC. The compression ratio is $r = 8$.

- Estimate the temperature, pressure and density at TC using the adiabatic
  relationships.
- Applying Bernoulli's equation between the plenum of the nozzle and the
  nozzle exit, estimate the exit velocity of the fuel from the nozzle. The
  density of the fuel is 0.74×10^3 kg/m^3.
- If the nozzle diameter is 20 μm, estimate the Reynolds number of the nozzle
  flow. Does this justify the application of Bernoulli's equation?

7. The entering fuel stream of the previous problem almost immediately breaks
up into droplets of diameter $d = 20\,\mu$m. We want to estimate the penetration
distance, the distance the droplet travels before it stops.
   At Reynolds numbers between 10 and 10^3 the (turbulent) drag coefficient
C_D of a fuel droplet varies between 4 and 0.4 [39]. For very crude estimates,
we may approximate this as a constant value of $C_D = 1$. Assume the droplet
enters at a speed of 106 m/s. We may assume that the droplet has arrived
at its stopping point when the Reynolds number has dropped to 10 (the
distance traveled after that point is less than 1 mm). The air in the cylinder
is in turbulent motion, but we can make an estimate of the upper limit of
penetration by taking the air to be still. To avoid confusion, we will designate
with a subscript d, for droplet, properties of the fuel droplet, and a subscript
g, for gas, properties of the gas in the cylinder. The droplet will obey the
equation

$$m_d\frac{du}{dt} = -C_D\frac{1}{2}\rho_g u^2 A_d \tag{5.45}$$

where m_d is the mass of the droplet,

$$m_d = \rho_d\frac{4}{3}\pi\frac{d^3}{8}, \tag{5.46}$$

ρ_d is the density of the droplet, A_d is the frontal area of the droplet,
$A_d = \pi d^2/4$, and ρ_g is the density of the gas. The equation can be reformed
as

$$\frac{d}{dt}\left(\frac{1}{u}\right) = \beta \tag{5.47}$$

where we have collected the various constants into

$$\beta = C_D \frac{3}{4} \frac{\rho_g}{\rho_d} \frac{1}{d} \qquad (5.48)$$

- You will need the kinematic viscosity ν_g of the gas in the cylinder. Recall that ν_g is proportional to $\sqrt{T_g}/\rho_g$.

- Solving the equation, obtain an expression for the particle velocity as a function of time.

- Integrate this to obtain an expression for the particle penetration as a function of time.

- Converting the expression for velocity into an expression for Reynolds number, evaluate the time of flight (of the particle) by using the initial and final Reynolds numbers.

- Using this number, determine the penetration distance.

We are neglecting a number of effects, notably the evaporation of the fuel droplet, which continuously reduces its size. This will reduce the penetration even more.

8. • Consider a single cylinder of an engine with given cylinder displacements and fixed compression ratio of 10.0. Assume that the combustion chamber shape is approximated by a right circular cylinder of radius $b/2$ with a flat top and flat base defined by the top of the piston. If there is a 0.02 inch thick quench layer (the air/fuel mixture does not burn at a distance of 0.02 inches from the chamber walls) at the walls exluding the flat base, determine the unburned/burned fuel ratio for the four cases of a bore/stroke ratio of 0.9 and 1.2, each for cylinder displacements of 100 cc (small motorcycle engine) and 500 cc (large automobile engine).

- Comment on the effects of the bore/stroke and displacement on pollution due to the quench layer.

- Comment on the effects of the variables on indicated power and efficiency.

9. A higher turbulence level at ignition causes a faster burn. The effect of this can be explored with ESP. Although ESP cannot simulate the tumble that produces the turbulence, the *factor in turbulence production during compression* which appears in the *Turbulence model* can be adjusted to increase the turbulence level at ignition to approximately twice the mean piston speed. Do this, and document the changes in the indicated efficiency, volumetric efficiency, cycle peak pressure, crank angle at end of burn, and *bsfc*.

6 OVERALL ENGINE PERFORMANCE

6.1
INTRODUCTION

We usually think of performance as the maximum torque and power available at any engine speed. However, except on an interstate on-ramp, we seldom use the maximum torque and power. Most of the time we are much more concerned with the behavior of the engine at partial throttle, transient response, and specific fuel consumption. Engine performance is strongly affected by the way in which the fuel is introduced.

6.2
CARBURETION VS. INJECTION

Despite the use of Gasoline Direct Injection in about 1932 by Hesselman, until the early 1970s most cars were carburetted. The carburetor has a narrowed throat, or venturi, in which the speed of the air entering the engine increases, causing the pressure to drop. The pressure drop between this point and the atmosphere causes the fuel to flow from the float chamber of the carburetor into the air stream. The fuel entering the airstream enters as liquid droplets. The mass of fuel entering must be controlled so that it is proportional to the mass of air flowing through the venturi. In addition, the droplet size must be controlled, the droplets must be encouraged to evaporate, the resulting vapor must be mixed uniformly with the air, and some source of heat must be provided to prevent the cooling resulting from the droplet evaporation from causing freezing of the water vapor in the air. In addition, provision must be made for modification of the mixture in response to engine temperature and operation (warm-up, idle, cruise, full power or acceleration). The modern carburetor is a fairly sophisticated device, with many small passages and jets, designed to meet all these needs under changing conditions. The carburetor can also be electronically controlled. Probably the culmination of the mechanically controlled carburetor is a carburetor for a military aircraft piston engine of World War II vintage, which must operate in all positions and altitudes.

However, the fuel vaporization and mixing in a carburetted engine is far from perfect even under the best circumstances. Fuel delivered to the various cylinders is not equal, some cylinders running a little rich and some a little lean.

Heywood [47] estimates cylinder-to-cylinder variation of ϕ from the mean of 5% at light load, 15% at WOT, and 20–30% for some engines at some speeds. Typically, the mixture is adjusted so that the leanest cylinder is rich enough to run reliably, which means that the richest cylinder is probably putting considerable unburned hydrocarbons into the exhaust. In addition, liquid fuel is usually deposited on the inside surface of the intake manifold, puddling on the lower surface, from which it evaporates. Under acceleration, the additional fuel sprayed from the accelerator jet is largely unevaporated.

6.2.1 Fuel Injection

In a successful effort to deal with some of these shortcomings in the face of economic and political pressure to reduce fuel consumption nationally, as well as to reduce pollution, particularly unburned hydrocarbons and oxides of nitrogen (that contribute to photochemical smog), fuel injection has been almost universally adopted (as well as electronic engine control). Throttle-body fuel injection offers the possibility of controlling the amount of fuel introduced more precisely, as well as reducing the droplet size somewhat, but still suffers from some of the problems of the carburetor with regard to distribution. Port fuel injection (PFI) has now become more-or-less standard. This overcomes the distribution problem so that all cylinders get a more nearly equal charge. The injection can be continuous or pulsed. Liquid fuel is still deposited on surfaces, here the back of the intake valve and the surrounding port walls. Because these are fairly hot, the fuel evaporates rapidly. In addition, under partial load there is backflow of residual exhaust gases into the intake manifold and these gases are hot; this increases evaporation. Nevertheless, [47] reports that some fuel is carried into the cylinder as liquid drops even when the engine is fully warm. Atomization in PFI is improved, since injection nozzles spraying fuel under a pressure of some 250–300 kPa can be designed to perform much better than a carburetor, where the pressure differential is less than 10 kPa.

Gasoline Direct Injection requires injectors operating at pressures of 5–7 MPa. Here it is necessary to form very small droplets, since it is desired to avoid having the spray impinge on the cylinder wall at all. Evaporation is achieved without the addition of heat from the valve, port or manifold.

6.2.2 Mixing and Evaporation

The air entering the throttle body, after passing through the air cleaner, is highly disturbed, and the flow will be turbulent. The flow will initially consist of the confused turbulent wakes of upstream objects, and the forming turbulent boundary layers on the (rough) walls of the manifold. Turbulence levels (that is, the ratio of r.m.s fluctuating velocity to mean flow velocity) in wakes of supporting struts, a couple of manifold diameters downstream are 30–50% [96]. Turbulence levels in turbulent boundary layers are of the order of 3–4% [96]. In a carburetted or PFI engine at partial throttle, the flow will be dominated by the mixing layers formed between the incoming flow and the separated region behind the throttle plate. Such regions typically have turbulence levels of about 25% [96]. Let us take levels of 25% as representative, and length scales of $D/6$, where D is the diameter of the throttle body throat or manifold branch, just to get a rough estimate of how long it will take to transport the fuel spray across the manifold (see Section 5.4, particularly Equation 5.12). In a distance x, in a pipe of diameter D, with a turbulence of intensity u and scale ℓ, in a mean velocity U, the turbulence can transport a scalar a transverse distance L of

$$L = \sqrt{\frac{2}{3}\frac{u}{U}\frac{\ell}{D}UtD}$$

$$= \sqrt{\frac{2}{3}\frac{u}{U}\frac{\ell}{D}xD} \tag{6.1}$$

or,

$$\frac{L}{D} = \sqrt{\frac{2}{3}\frac{u}{U}\frac{\ell}{D}\frac{x}{D}} \tag{6.2}$$

In $x = 25\,D$ this works out to be only $L/D = 0.8$. In $x = 10\,D$, we obtain $L/D = 0.5$. Ignoring the difference between transport of a droplet and transport of a gaseous scalar (which is not great for small enough droplets; in any event, this is a conservative answer, because droplets will not be transported as far as the gaseous scalar), it is clear that we do not have enough time to transport the fuel spray all across the manifold; that is, to thoroughly mix the fuel spray with the incoming air.

6.2.3 Droplet Size

Fuel sprays are typically rather coarse. A carburetor, operating at pressure differences in the neighborhood of 10 kPa, would produce droplet sizes in the neighborhood of several hundred μm. (Of course, any fuel spray contains a wide range of droplet sizes. Typically maximum sizes are between two and five times the mean size, depending on the nozzle design. I am giving here the mean diameter.) At this low pressure, however, the fuel is unlikely to be well-atomized, and probably includes few spherical drops, but rather mostly ligaments (connected strands of varying cross-section). Current understanding suggests that under most operating conditions (small throttle openings) these ligaments strike the throttle plate to form a film. The air flow past the edge of the throttle plate involves pressure ratios up to 2:1, and is nearly sonic; it will produce secondary atomization of the film as it leaves the edge of the plate, resulting in drop sizes in the neighborhood of 50 μm.

Throttle body or port fuel injection, operating at a pressure differential of perhaps 250 kPa produces droplet sizes of the order of 90 μm. Roughly speaking, the droplet size produced is inversely proportional to the square root of the pressure difference [107].

Droplets substantially larger than 10 μm (say, larger than 25 μm) cannot follow the air in the manifold around the corners, and will land on the walls [47]. If they remained in the airstream, they would not have time to evaporate. One problem is the decrease in droplet temperature – as the droplet begins to evaporate, its temperature drops rapidly (up to about 30°C), which reduces the vapor pressure [47]. Droplets originating at the throttle body (either from fuel injection or from carburetion), that are small enough to follow the flow (i.e., $\leq$10 μm) will evaporate before reaching the cylinder [47]. Hence, droplets will either land on the walls or will evaporate. The vast majority of the droplets are too large to follow the flow or to evaporate, and hence will land on the walls.

With port fuel injection, liquid fuel enters the cylinder during the intake stroke, and droplets persist through the compression stroke. In measurements reported in [47] the average droplet size during intake was of order 10–20 μm diameter; during compression, the average droplet size *increased*, because the smaller droplets were evaporating. At the end of injection, some 10–20% of the fuel was present in droplet form, but at ignition the fraction of the fuel in droplet form was negligible [47].

6.2.4 Puddling

I am including here Figure 6.1 from [47]. This figure shows the region of engine load and speed in which liquid fuel was observed in the manifold plenum and runner.

Once on the wall, the fuel forms a liquid film which evaporates into the airstream. The evaporation of the fuel involves a substantial amount of heat. A 1.5 L engine at 6,000 rpm requires approximately 2.6 kW to evaporate the fuel. This is the power consumption of a couple of very large refrigerators. There is plenty of wasted heat coming from the engine, but if it is not specifically provided to the

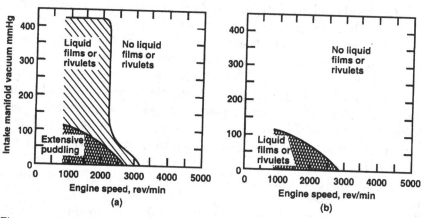

Figure 6.1. Regions of engine load and speed range where extensive pools or puddles, liquid films or rivulets were observed: (a) on manifold plenum floor, and (b) in manifold runner. Four-cylinder carburetted engine. Manifold heated by coolant to 90°C. From [56]. Reprinted with permission from 780944 © 1978 Society of Automotive Engineers, Inc.

manifold in a carburetted or throttle-body fuel-injected engine, the evaporating fuel will freeze the water vapor in the incoming air, and the engine will come to a stop when the throttle body is clogged with ice. In port fuel injection, the intake valve and port walls are warm enough to provide the heat, and icing is not a problem.

6.3
TRANSIENT RESPONSE

The existence of a liquid fuel film on the manifold wall means that there will be a delay in the engine response to changing throttle conditions. There is in any event a delay if the fuel enters at the throttle body, since (even if the air flow were incompressible) it would take a certain time for the changed amount of fuel to reach the cylinder. However, this is only of the order of one revolution of the engine. When conditions change, however, the film on the wall must be re-established at its new equilibrium thickness and extent. This is determined by a balance between removal of liquid fuel from the wall as droplets and evaporation of fuel from the wall due to the mass transport of the turbulent boundary layer over the surface, and the deposition of fuel droplets on the wall due to inertia. If the film is taken to cover a distance inside the manifold of 20 cm and to be 100 μm thick, then a time delay can be estimated (the time required to replace the film at the current $\dot{m}_f$) as something of the order of 30 revolutions. If the fuel film is thicker, it will take longer in proportion; e.g., at a thickness of 500 μm it will take 150 revolutions. More sophisticated calculations, reported in [47] agree with the order of this figure. Heywood [47] estimates the volume of fuel in the puddle of a 5 L V8 engine at part throttle as 1000 mm^3, a startling amount, corresponding to a puddle several mm deep and a substantial fraction of a meter long. During the time for the puddle to come to equilibrium again, if the throttle opening has been increased, the mixture is lean (since the new, increased concentration of fuel has not yet arrived at the cylinder). This is the purpose of the accelerator jet in the carburetor and of fuel enrichment in fuel injection, to provide extra fuel during this lean period.

6.4
BRAKE SPECIFIC FUEL CONSUMPTION

Brake specific fuel consumption is the rate of consumption of fuel per unit time and per unit power output. The traditional Imperial units were lb$_m$/HPh; in SI units, 1 lb$_m$/HPh = 0.608 kg/kWh. The brake specific fuel consumption is abbreviated bsfc. Recall Equation 1.12, Section 1.9, in Chapter 1:

$$P_b = \eta_m \eta_i \eta_c \dot{m}_f Q \tag{6.3}$$

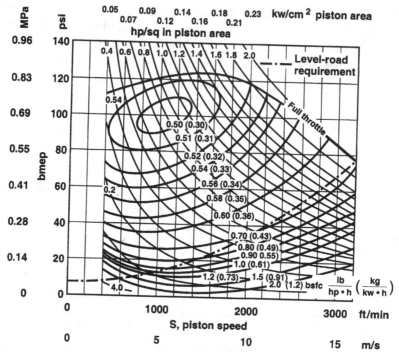

Figure 6.2. Performance map of a typical carburetted U.S. passenger-car engine as installed. Estimate of C. F. Taylor from accumulated test data [94]. Copyright © 1968 and new material © 1985 by The Massachusetts Institute of Technology.

This can be re-written as

$$\frac{\dot{m}_f}{P_b} = \frac{1}{\eta_m \eta_i \eta_c Q} \tag{6.4}$$

The left-hand side is the *bsfc*. For an engine operating on a given fuel, it is clear that the *bsfc* is a measure of the inverse of the product of the efficiencies. If one or more of the efficiencies goes up, the *bsfc* goes down, and *vice versa*.

Figure 6.2 shows the performance map of a typical carburetted US passenger car engine, as installed. The lines of constant specific power are obtained from

$$bmep = \frac{2X}{\overline{V}_p} \left(\frac{P_b}{A_p n} \right) \tag{6.5}$$

so that, for a fixed value of $P_b/A_p n$, *bmep* is proportional to the inverse of $\overline{V}_p$. The significant parts are the lines of constant *bsfc*, corresponding to lines of constant overall efficiency, the full throttle cut-off and the level road requirement.

Note the region of lowest *bsfc* (highest efficiency) at a relatively low piston speed and a relatively high *bmep*. This is where the product of indicated, mechanical and combustion efficiencies is highest.

Moving from this region of lowest *bsfc* on a line of constant piston speed, toward lower values of *bmep*, mechanical efficiency falls, because *imep* decreases while *fmep* remains nearly the same. Moving from the region of lowest *bsfc* on a line of constant piston speed toward higher values of *bmep*, the fuel:air ratio increases; the mechanical efficiency is also increasing, but the combustion efficiency is going down faster (due to the fuel:air ratio increasing), causing an increase in *bsfc*. The point here is, that although the fraction of fuel burned is going down, the absolute amount burned is still rising, so that the power is going up, and hence also the *bmep*.

Moving from the region of lowest *bsfc* along a line of constant *bmep* toward higher piston speed, *fmep* rises, while η_i remains nearly constant. This means a drop in η_m, and a rise in *bsfc*. Moving in the other direction from the minimum of *bsfc* on a line of constant *bmep*, η_c drops because of poor fuel distribution (in a carburetted engine) and increased relative heat loses, which cause a drop in η_i. Although *fmep* is decreasing, it is not enough, and *bsfc* rises.

Note the typical curve of *bmep* required versus piston speed for level road operation in high gear. As passenger automobiles are now designed, it is not possible to run on a level road at constant speed anywhere close to the regime of best fuel economy [94]. To get closer to the *bsfc* minimum it would be necessary to reduce the final drive ratio substantially – that is, reduce the engine speed at the same wheel speed. This is what an overdrive does, but a normal overdrive does not bring the operating point very near the *bsfc* minimum. A gear ratio that would move the operating point far enough would not allow any acceleration or hill-climbing ability, or permit very high speed. Automatic transmissions, which allow automatic kick-down into a lower gear when the accelerator is depressed, would permit an ultra-overdrive top gear ratio, with passing, hill-climbing and high speed performance still available on demand. Something like this is probably in our future, considering the pressure to reduce overall fuel consumption.

It is worth mentioning that in commercial aircraft, since the financial viability of the operation is absolutely dependent on fuel economy, the point of best *bsfc* is chosen for the cruising regime [94]. A piston aircraft engine will have a performance map quite similar to Figure 6.2. In piston-engined, propellor driven aircraft, variable pitch propellors make it possible for the point of minimum *bsfc* to correspond to cruise. On take off, the propeller pitch is reduced, and the fuel:air ratio substantially increased to give high take-off power; the sharply reduced *bsfc* is not important, since the time involved is short.

6.4.1 Power and Torque Curves

Usually the reader does not have access to a performance map for a given engine. At the most he is provided with a curve of maximum torque vs. engine speed, and/or maximum power versus engine speed. These are both equivalent to the full throttle curve of *bmep* vs. piston speed in Figure 6.2, because starting with

$$P_b = bmep \, A_p nS \, \frac{N}{X} \tag{6.6}$$

and writing

$$T = \frac{P_b}{2\pi N} \tag{6.7}$$

the torque, we can immediately write

$$T = bmep \frac{V_d}{2\pi X} \tag{6.8}$$

where V_d is the displacement. Hence, torque and $bmep$ are proportional for a given engine, as are piston speed and engine speed. Power is just torque times engine angular velocity:

$$P_b = 2\pi NT \tag{6.9}$$

Thus, power vs. engine speed can easily be obtained from torque vs. engine speed or from $bmep$ vs. piston speed.

Recalling Equation 1.17

$$\eta_i \eta_m \eta_v \eta_c \rho_i F Q = bmep \tag{6.10}$$

we can see that the $bmep$ at full throttle will fall at high piston speed because of the fall of η_m and η_v. It falls somewhat at low piston speeds because of the fall in η_c. Generally speaking, the curve of $bmep$ vs. piston speed is smooth and has a single maximum. An engine with tuned manifolds, particularly one with a switched manifold, or continuously variable valve timing, or valve timing that changes at a given speed in mid-range, may not have a smooth curve of $bmep$ vs. piston speed.

Let us give our attention to engines that have a smooth curve of torque vs. engine speed, with a single maximum. For such engines, the simplist data is usually given as the maximum brake power at the speed of maximum power, and the maximum torque at the speed of maximum torque, say P_b^m at $N_{P_b^m}$ and T^m at N_{T^m}. It is easy to construct crude curves from these numbers. We can write

$$P_b = 2\pi NT$$

$$\frac{dP_b}{dN} = \left(T + N\frac{dT}{dN}\right) 2\pi$$

$$\left.\frac{dP_b}{dN}\right|_{T^m} = 2\pi T^m \tag{6.11}$$

In exactly the same way, we can write

$$T = \frac{P_b}{2\pi N}$$

$$\frac{dT}{dN} = \frac{1}{2\pi N}\frac{dP_b}{dN} - \frac{P_b}{2\pi N^2}$$

$$\left.\frac{dT}{dN}\right|_{P_b^m} = -\frac{P_b^m}{2\pi N_{P_b^m}^2} \tag{6.12}$$

where I have made use of the fact that the slope of the torque curve is zero at its maximum, and the slope of the power curve is zero at *its* maximum. Using the first member of Equations 6.11 or 6.12, we can obtain the torque at the engine speed corresponding to maximum power, and the power at the engine speed corresponding to maximum torque. Hence, we have for each curve (power and torque) two values at two speeds, and the slopes of the curves at those values. Since we know that the curves are simple in shape, a French curve or spline fit will give a reasonable shape for the full curve.

6.5
PROBLEMS

1. Frictional losses in the valve train are approximately proportional to the loading, which is partly inertial and partly due to the valve spring. For a valve train that bounces at 20 m/s, the $fmep_v$ for the valve train can be written approximately as (in Pascals)

$$fmep_v = 20 \times 10^3 \tag{6.13}$$

A Desmodromic valve train has only inertial loading and no spring loading. As a result the $fmep_v$ for the desmodromic valve train can be written approximately as (in Pascals)

$$fmep_v = 3.77 \times 10^2 \, \overline{V}_p \tag{6.14}$$

where $\overline{V}_p$ is in m/s.
A desmodromic valve train might be expected to reduce the *bsfc*.

- You decide to modify your sports car by installing a desmodromic valve train. Determine the reduction in the total *fmep*, at piston speeds of 3 m/s, 12.5 m/s and 20 m/s, taking the total *fmep* to be that of Figure 4.5. Assume that $\overline{V}_p = 20$ m/s at 4,000 rpm.
- Using values for η_m from Figure 4.11, determine the change in η_m at these piston speeds for WOT.
- Then, using Equation 6.4, determine the percentage change in the *bsfc* at these piston speeds for WOT.

2. Consider the power equation for an engine. Consider possible four-stroke engines of fixed displacement running at fixed speed, with a fixed mixture and fuel of fixed heating value. Otherwise you, as designer, are free to select the parameters of the engine.

- What is the principal influence on η_i? Suggest one thing that can be done (when designing the engine) to keep η_i up.
- What is the principal factor controlling η_m? Suggest one thing that can be done (when designing the engine) to keep η_m up.
- What is the principal factor controlling η_v? Suggest one thing that can be done (when designing the engine) to keep η_v up.

7

DESIGN CONSIDERATIONS

7.1

INTRODUCTION

Why do you suppose that Cadillac introduced first a V12 engine, and then a V16 engine, back in the early 1930s? They were closely followed by Marmon, with its own V16, while Lincoln had a V12. Was this simply a marketing question: large numbers of cylinders had a certain mystique that was attractive to the customers? Or was there more to it than that? V12 engines have appeared elsewhere from time to time – Daimler in the UK had what they called a Double Six, and Jaguar introduced a V12 for the E-Type.

Why did the stroke:bore ratio change progressively over time? Just fashion, or something more?

These are design questions. They have to do with the descisions that a designer must make when a new engine is planned. Certainly marketing considerations come into it. There is no question that manufacturers have recently decided on four valves per cylinder in large part because the customers expect it; a Porsche with two valves per cylinder would sound wimpy. There is no doubt that Bentley and William Lyons (Jaguar) felt that double overhead cams had great market appeal, quite aside from their technical advantages. However, there are good technical reasons for some of these choices. Some of these we have mentioned in passing. Here, we will look into this in greater detail.

7.2

SIMILARITY CONSIDERATIONS

It is possible to examine some of these questions by considering geometrically similar engines of different sizes, or geometrically similar cylinders of different sizes. Two objects are geometrically similar if all dimensions are in proportion. Two engines, in one of which every dimension is one-half the value of the corresponding dimension in the other, would be geometrically similar. It is useful to consider an imaginary family of (geometrically) similar engines in a range of sizes, or an imaginary family of (geometrically) similar cylinders in a range of sizes.

Let us look at the question of the V16 vs. the V12, from a similarity point of view. Suppose that the piston speed, the *bmep* and the brake power of the two engines are all the same. Presume that all elements of a V16 cylinder (piston, connecting rod, crank throw, valves, etc.) are geometrically similar to the V12, but scaled down. How much do you expect the mass of the V16 block to be, relative to that of the V12? What will be the relative displacement? Suppose that the block temperature of both engines is maintained at 85°C. The V12 piston crown average temperature is 260°C at a piston speed of 14 m/s. At the same piston speed, what will be the piston crown average temperature in the V16?

If the *bmep*, the piston speed and the brake power are all the same, then the total piston area will have to be the same also (from Equation 1.8). Hence,

$$[A_p n]_{12} = [A_p n]_{16} \tag{7.1}$$

Since the piston area is proportional to the square of the bore, we have

$$[b^2 n]_{12} = [b^2 n]_{16} \tag{7.2}$$

or

$$\frac{b_{16}}{b_{12}} = \sqrt{\frac{12}{16}} = \sqrt{\frac{3}{4}} \tag{7.3}$$

If the V16 cylinder is geometrically similar to the V12 cylinder, its mass will be proportional to the cube of a linear dimension. It does not matter which one we use, since all are proportional. If we denote the mass of the block by M, we can write

$$\frac{M_{16}}{M_{12}} = \frac{[nb^3]_{16}}{[nb^3]_{12}} = \frac{16 b_{16}^3}{12 b_{12}^3} = \frac{4}{3}\left(\frac{b_{16}}{b_{12}}\right)^3 = \frac{4}{3}\left(\frac{3}{4}\right)^{\frac{3}{2}} = \sqrt{\frac{3}{4}} \tag{7.4}$$

Hence, we have found that the V16 block will have about 87% of the mass of the V12 block for the same power. In fact, the displacement will also be smaller. We can write

$$\frac{V_d^{16}}{V_d^{12}} = \frac{n A_p S_{16}}{n A_p S_{12}} = \frac{S_{16}}{S_{12}} = \frac{b_{16}}{b_{12}} = \sqrt{\frac{3}{4}} \tag{7.5}$$

So far as the piston crown average temperature is concerned, we need Equation 3.12, which we reproduce here:

$$\frac{T_p - T_c}{T_g - T_p} = \frac{A_{conv} St \rho c_p u L}{A_{cond} \gamma_{solid}} \tag{7.6}$$

On the right-hand side, the ratio of the convection and conduction areas will not change in geometrically similar cylinders. If the piston speed is the same in both engines, the turbulent velocity will be the same. Hence, the only thing that changes on the right-hand side is the conduction distance, L. We know that the average gas temperature $T_g = 683$ K. Hence, we can write

$$\frac{T_p^{16} - T_c}{T_g - T_p^{16}} = \frac{T_p^{12} - T_c}{T_g - T_p^{12}} \frac{L_{16}}{L_{12}} \tag{7.7}$$

Thus, we can write

$$\frac{T_p^{16} - T_c}{t_g - T_p^{16}} = \frac{533 - 358}{683 - 533}\sqrt{\frac{3}{4}} = 1.01 \tag{7.8}$$

which gives $T_p^{16} = 521\,\text{K} = 248°\text{C}$, a considerable reduction.

In this way, similarity considerations have shown us that the V16 with the same power and piston speed and *bmep* will be lighter, will have a smaller displacement, and will have lower piston crown temperatures. As an engine designer, you could now start asking somewhat different questions: evidently, with the same displacement and weight, at the same piston speed and *bmep*, we could have an engine of higher power, with the same piston crown temperatures.

In addition the V16 appealed to the market because of its unicity. It had disadvantages due to its complexity. These are the tradeoffs that a manufacturer must make.

7.2.1 Inertial Stress

The inertial stresses in the reciprocating parts are obviously a limiting factor, although as we shall discover in the Subsection 7.2.2, the inertial loading on the valve gear is probably a more serious limiting factor than the inertial loading on the connecting rod – piston assembly. In practice, the latter seldom fails in a well-designed engine conservatively operated, unless the connecting rod bearing has failed first, increasing the inertial loading substantially.

The acceleration of the reciprocating mass is proportional to the stroke and the square of the rotational speed:

$$\text{acceleration} \propto N^2 S \propto N^2 b \tag{7.9}$$

The mass of the reciprocating parts is proportional to the volume of these parts, which varies with the cube of a characteristic dimension, say the bore:

$$\text{mass} \propto b^3 \tag{7.10}$$

Hence, the force applied, say to the connecting rod, is proportional to the product of mass and acceleration:

$$\text{force} \propto b^3 N^2 b = b^4 N^2 \tag{7.11}$$

Finally, the stress is the ratio of force and cross-sectional area. The cross-sectional area is proportional to the square of a characteristic dimension, say the bore:

$$\text{cross-sectional area} \propto b^2 \tag{7.12}$$

Hence, the stress is given by

$$\text{stress} \propto \frac{b^4 N^2}{b^2} = b^2 N^2 \tag{7.13}$$

Since the mean piston speed is given by

$$\overline{V}_p = 2NS \propto 2Nb \tag{7.14}$$

among similar engines, we can identify Equation 7.13 as

$$\text{stress} \propto \overline{V}_p^2 \tag{7.15}$$

From a dimensional point of view the only thing we have left out is the density of the reciprocating parts. This would have given us the dimensionally correct form

$$\text{stress} \propto \rho \, \overline{V}_p^2 \tag{7.16}$$

Equation 7.16 is not much of an improvement on Equation 7.15 in comparing geo-metrically similar engines made of the same material. However, Equation 7.16 al-lows us to consider, for example, aluminum pistons versus cast iron pistons. It tells us that the allowable piston speed at which fracture will occur will be given by

$$\frac{\overline{V}_p^{iron}}{\overline{V}_p^{alum}} = \sqrt{\frac{\rho^{alum}}{\rho^{iron}}} \tag{7.17}$$

where $\overline{V}_p^{iron}$ is the critical speed for fracture of iron pistons, and $\overline{V}_p^{alum}$ the same for aluminum. Changing from cast iron pistons, at a density of 7.87 g/cc to aluminum pistons at a density of 2.70 g/cc thus permits an increase of the critical piston speed by a factor of about 1.7. This is very approximate, since we have to assume either that the connecting rod is made of the same material as the piston, or that the connecting rod is not contributing to the reciprocating load, neither of which is quite true.

The situation is more complicated if the connecting rod and the piston are made of dissimilar materials. In [99], the estimate is given that 1/3 of the connect-ing rod can be considered as oscillating. To make use of this, however, we need to know approximately the relative volume of the connecting rod and the piston. If they are taken as having equal volumes, the effective reciprocating density will be $\rho_{cr}/3 + \rho_p$, where ρ_{cr} is the density of the connecting rod, and ρ_p is the density of the piston. If we suppose that the connecting rod in both cases is made of forged steel, then the factor becomes (still supposing that the volumes of the piston and connecting rod are the same)

$$\frac{\overline{V}_p^{iron}}{\overline{V}_p^{alum}} = \sqrt{\frac{7.87/3 + 7.87}{7.87/3 + 2.70}} = 1.4 \tag{7.18}$$

7.2.2 Valve Speed

In modern engines, with stroke:bore ratios below unity, the inertial loading on the valve train is probably greater than that on the connecting rod. As a result, mecha-nical failure of the valve train will probably limit the top speed of the engine. We thus need to define a characteristic speed of the valve gear that will give us a crude measure of the stresses in the valve gear. Again, we can use similarity to do this.

Let us use D, the diameter of the valve head, as a characteristic dimension of the valve. The lift of the valve is a little more than one-quarter of the diameter, as we have noted:

$$\text{Lift} \propto D \tag{7.19}$$

As a result, the acceleration of the valve will be proportional to the square of the engine angular velocity and the lift:

$$\text{acceleration of valve} \propto N^2 D \tag{7.20}$$

The mass of the valve is proportional to the cube of the typical dimension:

$$\text{mass of valve} \propto D^3 \tag{7.21}$$

So the force on the valve is given by the product of mass and acceleration:

$$\text{force on valve} \propto N^2 D^4 \tag{7.22}$$

Finally, the stress is given by

$$\text{stress on valve} \propto \frac{N^2 D^4}{D^2} = N^2 D^2 \tag{7.23}$$

Using the expression for mean piston speed

$$\overline{V}_p = 2NS \tag{7.24}$$

allows us to write

$$\text{stress on valve} \propto \overline{V}_p^2 \frac{D^2}{S^2} \tag{7.25}$$

Now, if there are multiple intake or exhaust valves, say N_i intake valves, they will fill half the area of the cylinder. Hence, approximately

$$D^2 N_i \propto b^2 \tag{7.26}$$

so that finally, the stress will be proportional to

$$\text{stress on valve} \propto \overline{V}_p^2 \frac{b^2}{S^2} \frac{1}{N_i} \tag{7.27}$$

where N_i is either the number of intake valves, or the number of exhaust valves, whichever is smaller. That is, if there is only one exhaust valve, it will be larger, and have higher stresses, and hence it will be the limiting factor. Now, rather than use something proportional to a stress, it is more convenient to use a characteristic speed, since we are accustomed to working with the piston speed. Hence, take the square root, and call it a characteristic valve speed $\overline{V}_v$:

$$\overline{V}_v = \overline{V}_p \frac{b}{S} \frac{1}{\sqrt{N_i}} \tag{7.28}$$

The valve gear characteristic speed was first defined by Taylor [94]. It has the advantage that it involves only quantities that are easily found in the literature; valve diameters are difficult to find. However, it does not distinguish between very different designs. Overhead valves with pushrods involve quite different masses than overhead cam designs, and our assumption that the masses of both can be parameterized by the cube of the head diameter is obviously ridiculous. It is also clear that flathead and hemi- or penta-head designs will have valve diameters which are quite different fractions of the bore diameter. This, at least, could be removed by using the valve head diameter D rather than putting it in terms of the bore. However, it is a simple quantity that is nevertheless useful. The values tend to be about 15 m/s = 3000 ft/min, and there is no descernable trend with time, or difference between countries. Taylor [94] reports average figures for the US in 1965 of 16.4 m/s and in 1983 of 14.8 m/s; for imports in 1965, 12.7 m/s and in 1983 14.0 m/s. The range is from a low of 10.8 m/s to a high of 19.9 m/s. As expected, overhead cam engines tend to have higher values, since the masses involved are smaller, although the 1983 Jaguar E-type only has a value of 16.4 m/s. Taylor [94] reports values as high as 21.6 m/s for well-designed OHC racing engines.

7.2.3 The MIT Engines

In the years following the Second World War (reported in 1950, [93]), the Sloan Laboratory at MIT constructed three geometrically similar single-cylinder engines, with bores of 2.5 in, 4.0 in and 6.0 in. The engines were built of the same materials and had the same valve timing. The concept of geometric similarity is now so familiar that the effort of actually building similar engines might seem unnecessary. At the time, however, it was not so familiar, and the construction of such engines was probably essential to convince engineers working in industry of the utility of the concept. More important, however: we have been applying the idea of similarity to simple questions of inertial loading and the like. Similarity of the gas flow is a more involved question, and this justified the construction of the engines.

I include here (Figure 7.1) a drawing of the MIT engine, which was also used as the cover drawing for Taylor's book [94].

Using these engines, Taylor demonstrated that "Similar engines running at the same values of mean piston speed and at the same inlet and exhaust pressures, inlet temperature, coolant temperature, and fuel-air ratio will have the same volumetric efficiency within measurable limits" [94]. He also showed that the indicator diagrams of the intake and exhaust strokes (the low-pressure part of the cycle) were identical.

From a fundamental point of view we can reason that at the same mean piston speed, the gas velocities will be the same, since the ratios of b/D will be equal in similar engines, and the flow coefficients will be the same. The turbulent transport coefficients for momentum and heat, however, are proportional to $u\ell$, where u is

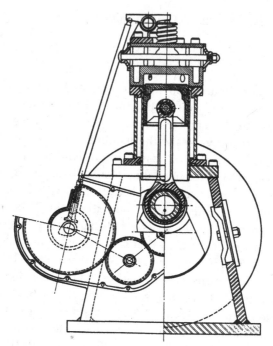

Figure 7.1. The MIT similar engine. This is the lateral cross section for all three engines, [93], [94]. Copyright © 1968 and new material © 1985 by The Massachusetts Institute of Technology.

a turbulent r.m.s fluctuating velocity and ℓ is a turbulent length scale for the energy-containing eddies. The turbulent velocity u will be proportional to the gas velocity, and hence will be the same in all the engines, but the length scale ℓ will be proportional to a typical geometrical scale, say b. Thus the transport coefficient will scale as

$$u\ell \propto b \tag{7.29}$$

Consequently, the flow cannot really be the same in the three engines.

The Reynolds numbers here are all large, well into the turbulent regime [96]. There is a small additional variation of the Stanton number and the drag coefficient with the Reynolds number, proportional approximately to the Reynolds number to the 0.2 power. The Reynolds numbers, based on the gas velocities, will be proportional to the bore:

$$R_e = \frac{UL}{\nu} \propto b \tag{7.30}$$

where U is a typical gas velocity (say, in the manifold), and L is a typical length, say a manifold length or diameter. The kinematic viscosity ν will be the same in all the engines since the same working fluid is used, and the same fuel, and the fuel:air ratio is the same. This suggests an additional variation in the Stanton number and the drag coefficient due to Reynolds number variation of approximately

10% between successive engines in the series. However, this is a small additional correction compared to the large difference in $u\ell$.

Fortunately, however, the turbulent pressure drop is an entirely negligible fraction (say, one part in 10^6) of the pressure drops occuring in the manifold (see Chapter 8); fluid inertia and separation at valve openings (another manifestation of inertia) are much more important. Hence, the difference in transport coefficients and Reynolds numbers is irrelevant, and the effects unmeasurable, so far as the pressure drop due to turbulent momentum transport is concerned.

We can make similar remarks about heat transfer. Part temperatures in the cylinder will certainly not be the same (as we have seen above) in different engines in the series. The temperature rise of the gas in the intake manifold, however, works out to be (if the Stanton number were constant) dependent only on dimensionless ratios such as u/U, the ratio of the turbulent fluctuating velocity to the gas mean velocity, and L/D, the length of the manifold relative to its diameter, which will be the same for similar engines. Hence, the only reason for a difference (among similar engines) of temperature rise for the gas in the manifold is the Reynolds number variation of the Stanton number. We have seen that this is of the order of 10%. The density of the gas at entry is proportional to the absolute temperature (at constant pressure). The temperature rise of the gas through the manifold is about 10% of the absolute temperature, in the neighborhood of 30°C. A change in the temperature rise of 10% is a change in the absolute temperature of 1%, and hence a change in the volumetric efficiency of 1%. That is within the experimental error.

Because the gas velocities are the same, the Mach numbers will be equal in the various engines. This is, of course, important because choking in the inlet valve flow is such an important factor in determining the volumetric efficiency.

7.3
BALANCE AND VIBRATION

The piston engine consists of pistons which reciprocate, a crankshaft that rotates and the connecting rod that does both. While it is possible to place counterweights on the crankshaft to balance the rotating forces, this is not possible for the reciprocating masses. We will give a brief introduction here to calculation of the forces and moments due to the reciprocating masses. This is an extensive subject, and it is fully treated in [94].

We must first deal with the connecting rod. We can replace the connecting rod with two masses, one at the wrist-pin end and one at the crank-pin end. The sum of these masses must equal the mass of the connecting rod, and the masses must be apportioned so that their mutual center of gravity is the same as that of the connecting rod. Then the mass at the wrist-pin end may be added to the mass of the piston to make a new reciprocating mass, and the mass at the crank-pin end can be added to the mass of the crank pin for a new rotating mass. In [99], a rule-of-thumb is given that the split of the two masses is approximately $1/3 - 2/3$, $1/3$ at the wrist-pin end.

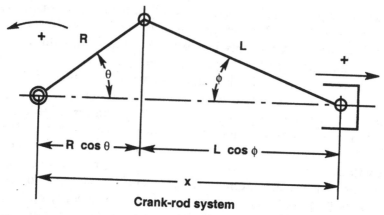

Crank-rod system

Figure 7.2. Crank-rod system [94]. Copyright © 1968 and new material © 1985 by The Massachusetts Institute of Technology.

We will assume that the counter weights on the crankshaft have been specified to balance both the crank-pin and the fraction of the connecting rod assigned to the crank-pin, nominally 2/3 of the connecting rod mass.

If the connecting rod were very long compared to the radius of the crank pin, this would all be much simpler, since the motion of the piston would be sinusoidal. Because the connecting rod is relatively short, there are higher order components in the motion. Refer to Figure 7.2.

The distance S of the wrist pin measured from the crankshaft center is given by

$$S = R \cos \theta + L \cos \phi \tag{7.31}$$

From the geometry, it is clear that

$$R \sin \theta = L \sin \phi \tag{7.32}$$

Using trigonometry, we obtain

$$\cos \phi = \sqrt{1 - \sin^2 \phi} = \sqrt{1 - \left(\frac{R}{L}\right)^2 \sin^2 \theta} \tag{7.33}$$

In practical engines, $R/L \leq 1/3$, so that $(R/L)^2 \leq 1/10$, roughly. This allows expansion of the expression in 7.33, keeping only the first term, which is most conveniently written in terms of the double angle:

$$\cos \phi \approx 1 - \frac{1}{2}\left(\frac{R}{L}\right)^2 \sin^2 \theta$$

$$= 1 - \frac{1}{2}\left(\frac{R}{L}\right)^2 \frac{1 - \cos 2\theta}{2}$$

$$= \left[1 - \frac{1}{4}\left(\frac{R}{L}\right)^2\right] + \frac{1}{4}\left(\frac{R}{L}\right)^2 \cos 2\theta \tag{7.34}$$

This allows us to write 7.31 as

$$S \approx R(a_0 + \cos\theta + a_2 \cos 2\theta) \tag{7.35}$$

where

$$a_0 \Rightarrow \frac{L}{R} - \frac{1}{4}\frac{R}{L} \tag{7.36}$$

and

$$a_2 = \frac{1}{4}\frac{R}{L} \tag{7.37}$$

Neglected terms are of the order of 1/100. Writing $\theta = \omega t$, we can obtain expressions for the velocity and the acceleration of the piston:

$$\dot{S} \approx -R\omega(\sin\theta + 2a_2 \sin 2\theta)$$
$$\ddot{S} \approx -R\omega^2(\cos\theta + 4a_2 \cos 2\theta) \tag{7.38}$$

Using Equation 7.38 we can write the force F_a acting on the engine structure, at the center of the crankshaft, as an (equal and opposite) reaction to the force required to accelerate the piston (plus the fraction of the connecting rod that oscillates). Let us call the mass of the piston plus this part of the connecting rod M_p.

$$F_a \approx M_p R\omega^2(\cos\theta + 4a_2 \cos 2\theta) \tag{7.39}$$

Note that $4a_2 = R/L$. Using Taylor's notation [94], we will call $M_p R\omega^2 = Z$ for convenience.

We can see from Equation 7.39 that there are two forces: the first order, $Z\cos\theta$ which rotates at the crankshaft speed, and the second order term $Z\frac{R}{L}\cos 2\theta$ which rotates at twice the crankshaft speed.

7.4
THE IN-LINE FOUR

7.4.1 The Forces

In a multi-cylinder engine, these forces must be added vectorially, aligned with the crank-pins. For example, consider a four-cylinder, four-stroke engine, with the cylinders uniformly spaced in line, axes parallel and the heads at the same end. All cylinders will fire in two revolutions, so that the firing interval is π. This means that the crankpins are separated by an angle of π, and the crankshaft is consequently flat. Let us call the first order force associated with cylinder 1 (the number is the firing order – 1 is the first cylinder to fire, etc.) F_{a1}^1, that associated with cylinder 2 F_{a2}^1 and so forth. Then, at a given instant, F_{a1}^1 is at an angle θ_1, F_{a2}^1 is at an angle $\theta_1 + \pi$, F_{a3}^1 is at an angle $\theta_1 + 2\pi$ and F_{a4}^1 is at an angle $\theta_1 + 3\pi$. Consequently F_{a1}^1 and F_{a3}^1 are aligned, and F_{a2}^1 and F_{a4}^1 are also aligned, rotated

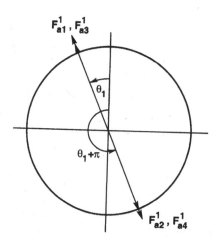

First order forces

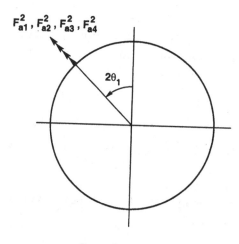

Second order forces

Figure 7.3. Graphical representation of the first and second order forces on a four-cylinder in-line, four-stroke engine.

by an angle π. The first order forces are therefore balanced. These forces are shown added vectorially in Figure 7.3. Note that in the graphical representation in Figure 7.3, involving vector addition of the forces, it is *only the components of the forces in the direction of the cylinder axes which are relevant.*

Let us call the second order force associated with cylinder 1 F_{a1}^2, and so forth. At the same instant, F_{a1}^2 is at an angle $2\theta_1$, F_{a2}^2 is at an angle $2\theta_1 + 2\pi$, F_{a3}^2 is at an angle $2\theta_1 + 4\pi$, and F_{a4}^2 is at an angle $2\theta_1 + 6\pi$. Consequently, all the second order forces are pointing in the same direction, and they result in an unbalanced force

$$4Z\frac{R}{L}\cos 2\theta \qquad (7.40)$$

These are also shown in Figure 7.3.

7.4.2 Moments

There is also the question of moments. The crankshaft is flat, with two throws up and two throws down. We have seen that the primary forces from cylinders 1 and 3 are in the same direction, and those from 2 and 4 are in the same direction. If we place the "up" throws at each end of the crank, and assign them cylinders 1 and 3, then the forces from these cylinders will not produce a moment about an axis at right angles to both the crank axis and the cylinder axis. Cylinders 2 and 4 will then take the middle two throws (which are both "down"), and again they cannot produce a moment. This determines the two classical firing orders for this engine: if the cylinders are numbered from the front of the engine (rather than by firing order) then the firing order is either 1243 or 1342, depending upon the order in which cylinders 2 and 4 (numbered by firing order) are assigned to the two middle throws. We show a side view of the crankshaft, with the first-order forces indicated, in Figure 7.4.

The second order forces cannot produce a moment either because they are all pointing in the same direction, one on each throw, so they produce no moment no matter how the cylinders are assigned to the throws. These are indicated in Figure 7.5.

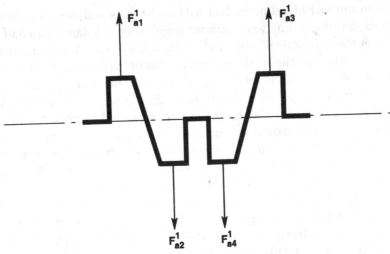

Figure 7.4. Sketch of the crank of a four-cylinder in-line, with the first-order forces indicated (side view), indicating that no moment is generated.

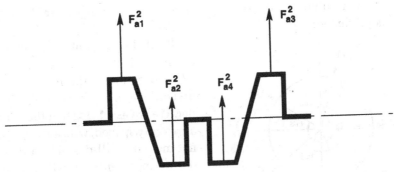

Figure 7.5. Sketch of the crank of a four-cylinder in-line with the second order forces indicated (side view), indicating that no moment is generated.

7.4.3 Balance Shafts

The unbalanced second order forces can be balanced by two balance shafts, symmetrically located relative to the plane containing the cylinder axes, rotating in opposite directions at twice the crankshaft speed, each carrying an unbalanced weight symmetrically located from front to back of the engine. The horizontal components of their unbalanced reaction will cancel, but the vertical components will add, phased to cancel the second order forces.

7.5
THE FIVE CYLINDER IN-LINE

While two-stroke engines could be made with any number of cylinders, four-stroke engines have traditionally been made with an even number of cylinders, primarily because moments can be made to cancel. Not that all four-cycle engines with an even number of cylinders had balanced forces and moments. We have already seen that the in-line four-cylinder engine has unbalanced second order forces. Uncounted millions of in-line fours have been built, beginning with the model-T Ford, and into the modern period, without bothering about balancing these second-order forces. The flat four, used in the VW Beetle, of which 15 million were manufactured, had balanced forces, but unbalanced moments. The 120° V6, which has become quite popular, has balanced forces, but unbalanced moments.

Balance shafts have made it possible to balance practically any unbalanced force or moment. However, many engines with unbalanced forces or moments do not have balance shafts. In particular, Volvo has come out with a five-cylinder in-line, which does not have balance shafts. Let us examine this engine.

The firing interval for this engine is $4\pi/5$. Hence, the first order forces for the five cylinders will be at angles of θ_1, $\theta_1 + 4\pi/5$, $\theta_1 + 8\pi/5$, $\theta_1 + 12\pi/5$, and $\theta_1 + 16\pi/5$. These are clearly uniformly spaced, and hence add to zero. The second order forces are at $2\theta_1$, $2\theta_1 + 8\pi/5$, $2\theta_1 + 16\pi/5$, $2\theta_1 + 24\pi/5$ and $2\theta_1 + 32\pi/5$. Again, these are clearly uniformly spaced, and hence cancel. We show the vector addition of these forces in Figure 7.6. The moments, however, are another matter. With an odd number of throws, there is no way to arrange a crank so that no moment is produced.

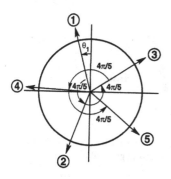

First order forces

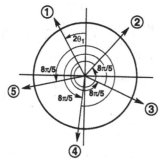

Second order forces

Figure 7.6. Vector addition of the first and second order forces on a five-cylinder in-line engine.

If all the possible firing orders are considered, there are 24 possibilities. We can compute the moments of the first order forces about the middle cylinder. We are assuming that the cylinders are evenly spaced, with a spacing d. The net moment will depend on the cylinder arrangement. We will continue to use the numbers 1, 2, etc. to indicate the first cylinder to fire, the second, and so forth. When we have selected a cylinder arrangement, we will translate that into the usual firing order.

The cylinder in front of the middle cylinder will have a moment of $+d$ times the first order force on that cylinder. The cylinder in front of that will have a moment of $+2d$ times the first order force. The cylinder behind the middle will have a moment of $-d$ times the first order force, and the cylinder behind that will have a moment of $-2d$ times the first order force. These are the forces shown in the top part of Figure 7.6. These must now be added vectorially, taking account of the signs and the multipliers ($\pm d$, $\pm 2d$). We will find that the vast majority of the vector diagrams produced by various cylinder arrangements have all the moment vectors on one side of the diagram, so that they add together, producing relatively large moments. Four diagrams are relatively balanced, and of these two have the smallest residual value. One of these corresponds to the cylinder order 54132, and the other to 23145, the reverse order. We show the first of these in the upper part of Figure 7.7. The net value is $0.45 Zd \sin \theta_1$ for the vertical component.

We are not so fortunate with the second order forces. There is a cylinder arrangement that gives a corresponding low value for the second order moment, but its corresponding first order moment is large. The second order moment of our arrangement 54132 is a whopping $-4.98 Zd\frac{R}{L}\sin 2\theta_1$ (see the lower part of Figure 7.7), resulting from all the vectors being on the same side of the diagram. We could pick a cylinder arrangement that gives a small second order moment, and a larger first order moment, but the second order moment is easier to isolate, since the frequency is double that of the first order moment. Hence, it makes more sense to keep the first order moment small, and put up with a larger second order moment.

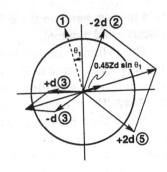

First order moments
Cylinder order 54132

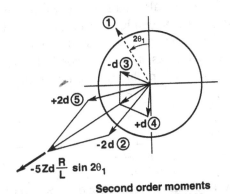

Second order moments
Cylinder order 54132

Figure 7.7.   The vector sum of the first and second order moments of the five-cylinder in-line engine, cylinder order 54132.

To put these values in context, the 120° V6 has a maximum vertical first order moment of $0.867 Zd \sin \theta_1$ [94]. One arrangement of straight eight has a second order moment of $4Zd \cos 2\theta_1$. The moments of the five-cylinder engine are of this order. Many engines live with moments of this order and make no attempt to cancel them with balance shafts, using isolation with soft engine mounts instead. The second order moment could be cancelled using two counter-rotating shafts, rotating at twice crankshaft speed. Each shaft would carry a pair of weights, one at each end, and diametrically opposite each other. Such a shaft would produce a rotating moment, and the horizontal components produced by the two shafts would cancel. A similar arrangement could be used to cancel the first order moments. It would probably be satisfactory to use the crankshaft as one balance shaft and run a single balance shaft as close to it as possible in the opposite direction. The only problem with this is the fact that it would not be possible to have this odd pair centered on the cylinder centerline, so there would be a small residual uncancelled moment.

Our cylinder arrangement 54132 corresponds to a firing order 35421 (now the numbers correspond to the position of the cylinders from the front of the engine). It is customary to give the firing order starting with cylinder 1, and this corresponds to 13542. The equivalent reverse order is 12453.

Volvo uses this firing order and hence has minimized the first order oscillating moment, depending on good vibration-isolation engine mounts to minimize the second order oscillating moment. The five-cylinder Volvo engine does not use balance shafts. The Volvo engineers point out that the first order oscillating moment is equivalent to two counter-rotating first order moments. Volvo cancels one of these (the one that rotates in the same direction as the crankshaft) by extra balance weights on the crankshaft, but leaves the other. This leaves the engine with an unbalanced first order rotating moment, rotating in a direction opposite to that of the crank. I do not understand why this should be better than having an unbalanced first order oscillating moment, or why it should be easier to isolate. However, it is at least no worse. The engine has an excellent reputation in every respect, so it probably is immaterial.

7.6

PROBLEMS

1. Given the following data on a particular car: Piston mass $= 0.54$ kg, $bmep = 1.196$ MPa, bore $b = 84$ mm, stroke $S = 104$ mm, connecting rod mass $= 0.47$ kg, compression ratio $= 8.0$.

   • Calculate the maximum gas force on the piston surface assuming that the maximum gas pressure is estimated as $= 2\ bmep$.

   • Calculate the inertial forces $=$ (mass)(accel.), of the oscillating piston-rod assembly at 3,000 and 6,000 rpm, assuming that the piston has approximately a sinusoidal displacement over time. Note that the connecting rod undergoes both translational and rotational motion.

   • Find the rpm where the gas force $=$ inertial force. $bmep$ here is the peak $bmep$ over the operating range of the engine.

2. Draw the vector diagrams for a four-cycle horizontally opposed two-cylinder engine with crank-pins 180° apart. This engine will have a firing interval of 2π. Show that first and second order forces cancel, but that first and second order moments do not.

3. Consider two engines, A and B. Engine A has a stroke $S_A = 2b_A$, where b_A is the bore diameter of engine A. Engine B has the same displacement, but $S_B = b_B$. In each engine the intake valve diameter is equal to the exhaust valve diameter, both equal to half the bore. The lift of each valve is 1/4 of the diameter. The engines are otherwise the same, and have the same

compression ratio and valve timing.

- Let N_A and N_B be the speed of engine A and B at which the Mach index is equal to 1.0. What is the value of the ratio N_B/N_A?

- Now let N_A and N_B be the speed of engines A and B at which the mechanical efficiency is 0.6. What is the ratio N_B/N_A?

- Supposing that the thickness of the piston crown is the same in engines A and B, what is the ratio $\Delta T_B/\Delta T_A$ of the temperature difference $\Delta T =$ [pistoncrown − cylinderwall] in the two engines at the same rpm? Suppose there is no oil spray on the underside of the crown, and that all heat is lost to the cylinder wall.

THE STANFORD ESP

8.1

INTRODUCTION

The Stanford Engine Simulation Program (ESP)[1] is a fast-running, flexible, user-friendly interactive program, designed to run on IBM compatible personal computers or mainframes, for simulation of the thermodynamic performance of homogeneous charge engines. It was developed at Stanford University for instructional purposes, but should also be useful to engine designers. A single cylinder is considered using a zero dimensional thermodynamic analysis, a simple geometrical approach to flame structure and a one-equation dynamical turbulence model that allows the effects of turbulence on heat transfer and combustion to be examined.

The user specifies geometric parameters of the engine, including bore, stroke, rod length, valve lift and timing, and heat transfer area above the piston at TC. The program handles both conventional crankshaft engines and engines with unequal compression and expansion strokes, and a variety of valve motion programs. The user also specifies operating parameters including engine speed, spark timing, valve timing, exhaust gas recirculation, and manifold pressures if no manifold model is included. When an intake or exhaust manifold is included, the user specifies the geometry of the manifold, information on flow restrictions (throttles), and friction parameters. These and other model parameters may be saved in an engine set-up file, which is automatically loaded when the user calls ESP. The parameters can be adjusted to get reasonable agreement with actual engine data, and the model then used to study the effects of proposed design changes.

ESP provides instructions and explanations to the user, and is easily learned without a manual. Executeable versions for Windows, UNIX, and other systems are available on the ESP website (http://esp.stanford.edu). This and supporting software are freeware and may be copied by anyone for any use. The Fortran source code is fully documented internally, and is available for licensed use through W. C. Reynolds.

---

[1] This chapter is based on a description prepared by W. C. Reynolds.

8.2
OUTLINE OF THE MODEL

To describe what happens inside the cylinder, and at the entrance to and exit from the cylinder, the model uses ordinary differential equations derived from energy balances, mass balances, and a turbulence model equation, as well as algebraic equations relating the variables. The gas in the cylinder is idealized as perfectly mixed except during the burn stage, in which case two zones (burned and unburned) are employed. Modeling of the motion of the gases in the manifold is separately described in Section 8.4, ESP Manifold Analysis.

In order to speed the calculation, thermodynamic properties of the fresh charge (reactants) and products are pre-calculated using a STANJAN-derivative program ESPJAN and loaded as tables into ESP. Product dissociation is considered, with thermodynamic equilibrium assumed. Since the equilibrium composition is a weak function of pressure, the table is constructed with a single average pressure for the reactants (e.g., 2 atm) and one for the products (e.g., 6 atm). The gas in ESP is treated as a mixture of these reactants and products, so the effects of residual gases and EGR are treated (the user specifies the direct EGR fraction).

The flow rates through the valves are computed using isentropic compressible flow theory with assumed discharge coefficients. In computations without manifolds, the intake and exhaust manifold pressures are taken as specified constants. Backflow through the intake valve is considered, with the backflow gas assumed to be homogeneous; it is assumed not to mix with the intake charge. The exhausted gas from each single cycle is assumed to be homogeneous in the cylinder and exhaust manifold, and backflow into the cylinder is allowed.

Heat transfer between the cylinder gases and the walls, at a set wall temperature, is allowed. The instantaneous rate is computed using a user-specified Stanton number based on the turbulence velocity. That is, no mean motion (swirl/tumble) is assumed for heat transfer in the cylinder. The heat transfer rate between the valve flow and specified heat transfer area is computed using a user-specified Stanton number based on the velocity through the valve.

Ignition is assumed to occur at a specified crank angle with the instantaneous burn of a small specified fraction of the unburned gas. The user specifies a geometric table giving the projected flame area (i.e., the area of a spherical surface centered on the spark plug and advancing at the effective turbulent flame speed) and wall heat transfer area behind the flame (behind this spherical surface), each divided by the bore area, as functions of the fraction of volume burned. This flame area, together with the evolving turbulence velocity and a specified laminar flame speed (both of which are used to determine an effective turbulent flame speed), are used to determine the

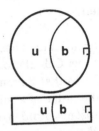

Figure 8.1. Cartoon of the burned and unburned zones in a simple pill-box combustion chamber, such as that in a CFR engine.

burn rate. This approach allows estimation of the effects of turbulence level, spark plug location, and combustion chamber geometry on engine performance.

The turbulence model is a unique feature of ESP, and is used to calculate the turbulence velocity parameter used in the flame speed and heat transfer models. The model computes the average turbulent kinetic energy per unit mass of the in-cylinder gas, again assuming homogeneity of the burned and unburned gases. It allows for kinetic energy inflow or outflow through the valves, production of turbulence kinetic energy due to shearing caused by piston motion and by density change (made consistent with rapid distortion theory), and dissipation of turbulence energy. Coefficients in this model are user-specified and can be adjusted to simulate different in-cylinder turbulence control techniques.

The calculation proceeds in four stages; *compression, burn, expansion,* and *gas exchange*. The compression stage begins when the intake valve closes and ends at ignition. The expansion stage begins at the end of burn and continues until the exhaust valve opens. If this happens before the burn is completed, the burn is quenched when the exhaust opens and there is no expansion stage. The gas exchange stage runs from exhaust-open to intake-close.

The solution requires integration of a few ordinary differential equations in combination with some algebraic equations. The integration uses a second-order Runge-Kutta method with time steps corresponding to one crank angle degree. Two first-order steps are taken whenever one stage ends and another begins between time steps. The calculation is structured so that the algebraic equations are maintained *exactly*; this is not the case in engine codes that differentiate the algebraic equations to get differential equations for all variables. Because of this special treatment, ESP can use a larger time step and hence run faster than other codes, especially during burn, with similar accuracy.

At the end of each cycle, a one-line summary of the convergence data for the cylinder is displayed. A one-line convergence display is also given for each manifold. Runs without manifolds converge very well in three or four cycles, which are virtually instantaneous on modern personal computers. Runs with manifolds take much longer. When happy with the convergence, the user can make MATLAB® plot files for the indicator diagram, manifold pressures and velocities, and selectable in-cylinder variables. These can beviewed immediately in an associated MATLAB® window. A summary of the performance data can be displayed, and a complete tabulation of the model constants used and the performance calculations can be written to a file.

The manifold calculations yield the pressure and velocity histories at key points in the manifold. These are very instructive and can be used for first-order manifold tuning. Calculations with an intake manifold, which is dominated by expansion waves that flatten, converge relatively quickly (say 20 cycles), which requires of the order of a minute on a modern personal computer. Calculations with an exhaust manifold take longer because of the large entropy variations in the manifold and steepening compressive waves. Calculations with both manifolds take even longer, typically several minutes per case. Therefore, a user should do as much as possible without manifolds before including them in the analysis.

ESP is significantly faster than zero-dimensional codes used in the industry but not quite as powerful as some (See, for example, some current proprietary codes [35], [82], as well as the slightly dated codes described in [70]. Also see http://www.sae.org, where references to some current work can be found). The best industrial engine codes employ more accurate integration methods, more accurate flame geometry models, do a chemical calculation at each time step, and include some reaction kinetics for emissions prediction. However, most do not incorporate a turbulence model, relying instead on empirical results that are very engine-specific and are not as flexible and user-friendly as ESP. Thus, ESP is expected to find use both in academic institutions and for preliminary analysis in the engine industry; its development will be continued with both user communities in mind.

8.3
MODEL DETAILS

8.3.1 Gas Properties

It is assumed that the internal energy u and enthalpy $h = u + Pv$ of the reactants and products are functions only of temperature. The user calls ESPJAN to calculate a table giving the values of h, and Pv as function of T at intervals of 100 K from 200 K to 4,900 K (48 points). The pressures to be used for these calculations (one for reactants, e.g., 2 atm, and one for products, e.g., 6 atm) are chosen by the user. These tables are saved to be read by ESP. A new table is required whenever the fuel, oxidizer, or their ratios are changed.

8.3.2 Analysis of the Compression Stages

During the compression stage, the energy balance on the in-cylinder gas is (see Figure ??)

$$\frac{dU_c}{dt} + \dot{W}_p + \dot{Q}_c = 0 \qquad (8.1)$$

where U_c is the total internal energy of the gas in the cylinder, $\dot{W}_p$ is the power output to the piston, and $\dot{Q}_c$ is the heat transfer rate from the gas to the head, cylinder walls, and piston. $\dot{W}_p$ is computed from

$$\dot{W}_p = P A_p \frac{dz}{dt} \qquad (8.2)$$

where P is the pressure, A_p the bore area, and z the piston distance from TC. $\dot{Q}_c$ is computed from

$$\dot{Q}_c = St_c V_t \rho c_p A_c (T_c - T_w) \qquad (8.3)$$

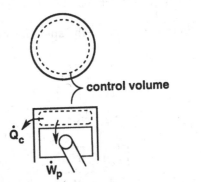

control volume

Figure 8.2. Cartoon of the energy balance on the in-cylinder gas.

Here St_c is a user-specified Stanton number based on the turbulence velocity V_t, the gas density ρ, and the gas specific heat at constant pressure c_p; A_c is the heat transfer area, T_c the gas temperature, and T_w is the user-specified wall temperature.

Equation (8.1) is used to advance U_c to the new time step. U_c is then divided by M_c, the gas mass in the cylinder, to get the specific internal energy u_c of the gas. Then, using the properties table, the new temperature T_c is determined, yielding u_c for the mixture of products and reactants currently in the cylinder. The specific volume v is calculated from the new cylinder volume and mass, and then the pressure is calculated from the value of Pv determined from the table for the gas mixture at T_c. The turbulent kinetic energy $k = V_t^2/2$ is advanced using the model equation for k (see below), and then the new V_t is calculated.

8.3.3 Ignition Analysis

Ignition is treated as an abrupt discontinuity between the compression and burn stages occurring at a user-specified crank angle with the instantaneous conversion of a specified mass fraction f of the reactants to products ($0 < f \leq 1$). This produces an unburned zone and a burned zone, each assumed to be homogeneous and hence characterized by a single state. The two zones are assumed to be at the same pressure, and the unburned gas is assumed to be compressed isentropically (no heat transfer and no piston work during this instantaneous conversion).

The mass balance is (see Figure 8.3)

$$M_u + M_b = M_c \tag{8.4}$$

where M_u and M_b are the masses of the unburned and burned zones after ignition, respectively. The energy balance is

$$M_u u_u(T_u) + M_b u_b(T_b) = U_c \tag{8.5}$$

where u_u and u_b are the specific internal energies and T_u and T_b are the corresponding temperatures of the two zones. Since the volume is constant during this instantaneous process,

$$M_u v_u(T_u, P_+) + M_b v_b(T_b, P_+) = V_c \tag{8.6}$$

where v_u and v_b are the specific volumes of the two zones and P_+ is the pressure after ignition. The isentropic compression of the unburned gas requires

$$s_u(T_u, P_+) = s_u(T_-, P_-) \tag{8.7}$$

where s_u denotes the specific entropy of the unburned gas, T_u is the unburned gas temperature after ignition, and P_- and T_- are the pressure and temperature in the cylinder before ignition.

The burned mass is calculated from the mass fraction specified for ignition,

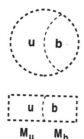

Figure 8.3. Cartoon of the mass balance in the cylinder.

and the unburned mass from the mass balance. Then, (8.5), (8.6), and (8.7) provide three simultaneous non-linear equations for P_+, T_u, and T_b. These are solved simultaneously by linearizing about a trial solution. Denoting the trial state by $\tilde{}$, the linearized energy balance (8.5) becomes

$$M_u[\tilde{u}_u + \tilde{c}_{vu}(T_u - \tilde{T}_u)] + M_b[\tilde{u}_b + \tilde{c}_{vb}(T_b - \tilde{T}_b)] = U_c \tag{8.8}$$

Multiplying the volume constraint (8.6) by P and linearizing,

$$M_u[\tilde{P}\tilde{v}_u + \tilde{R}_u(T_u - \tilde{T}_u)] + M_b[\tilde{P}\tilde{v}_b + \tilde{R}_b(T_b - \tilde{T}_b)] = \tilde{P}\,V_c \tag{8.9}$$

Since the unburned gas will only be slightly compressed by the ignition process (because f is small), we can treat it as having a constant specific heat for the isentropic compression process; then, the entropy condition gives

$$\frac{\tilde{P}_+}{P_-} = \left(\frac{\tilde{T}_u}{T_-}\right)^{\gamma/(\gamma-1)} \tag{8.10}$$

In this iteration, (8.10) is used to find $\tilde{P}_+$, and then (8.8, 8.9) are solved simultaneously for the temperature perturbations. Improved temperatures are then calculated, and the process is repeated to convergence.

8.3.4 Analysis of the Burn Stage

The burn stage analysis is based on energy and mass balances on the unburned and burned zones, each assumed to be homogeneous and hence characterized by a single state. The boundary between them is placed on the unburned side of the flame, and the pressure is assumed to be the same in both zones.

The energy balance on the unburned zone is

$$0 = \dot{M}_b h_u + \dot{Q}_u + P_c\frac{dV_u}{dt} + \frac{dU_u}{dt} \tag{8.11}$$

where $\dot{M}_b$ is the mass transfer rate to the burned zone; $\dot{Q}_u$ is the heat transfer rate from the unburned zone to the walls; P_c is the pressure in the cylinder; V_u the unburned volume; and U_u is the total internal energy in the unburned zone. This is shown symbolically in Figure 8.4. The energy balance for the burned zone is

$$\dot{M}_b h_u = \dot{Q}_b + P_c\frac{dV_b}{dt} + \frac{dU_b}{dt} \tag{8.12}$$

where $\dot{Q}_b$ is the heat transfer rate from

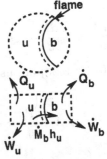

Figure 8.4. Cartoon of the energy balance in the burned and unburned zones with the various terms in Equations 8.11 and 8.12 identified.

the burned zone to the walls; V_b and U_b are the volume and internal energy of the burned zone. The mass balances are

$$0 = \dot{M}_b + \frac{dM_u}{dt} \tag{8.13}$$

and

$$\dot{M}_b = \frac{dM_b}{dt} \tag{8.14}$$

where M_u and M_b are the zone masses.

The volume provides a constraint on the solution,

$$V_c = V_u + V_b = \frac{M_u R_u T_u}{P_c} + \frac{M_b R_b T_b}{P_c} \tag{8.15}$$

Some manipulations are necessary to set up an effective way to enforce this constraint on the solution. Multiplying by P_c and then differentiating, and using the mass balances, (8.15) becomes

$$V_c \frac{dP_c}{dt} + P_c \frac{dV_c}{dt} = M_u \alpha'_u \frac{dT_u}{dt} + M_b \alpha'_b \frac{dT_b}{dt} + \dot{M}_b(\alpha_b - \alpha_u) \tag{8.16}$$

where $\alpha(T) = RT$ and $\alpha' = d\alpha/dT$. We will use this momentarily.

It is useful to rewrite the energy balance on each zone in terms of its total enthalpy $H = U + PV$,

$$0 = \dot{M}_b h_u + \dot{Q}_u - V_u \frac{dP_c}{dt} + \frac{dH_u}{dt} \tag{8.17}$$

$$\dot{M}_b h_u = \dot{Q}_b - V_b \frac{dP_c}{dt} + \frac{dH_b}{dt} \tag{8.18}$$

Now, expressing the zone enthalpies as $H = Mh$, expanding, and using the ideal gas equation of state $dh = c_p dT$, the energy balances give

$$M_u c_{pu} \frac{dT_u}{dt} = V_u \frac{dP_c}{dt} - \dot{Q}_u \tag{8.19}$$

and

$$M_b c_{pb} \frac{dT_b}{dt} = V_b \frac{dP_c}{dt} - \dot{Q}_b + \dot{M}_b(h_u - h_b) \tag{8.20}$$

When (8.19, 8.20) are used to replace the temperature derivatives in (8.16), one finds

$$(V_c - \beta_u V_u - \beta_b V_b) \frac{dP_c}{dt} = -P_c \frac{dV_c}{dt} + \beta_b \dot{M}_b(h_u - h_b)$$
$$- (\beta_u \dot{Q}_u + \beta_b \dot{Q}_b) + \dot{M}_b(\alpha_b - \alpha_u) \tag{8.21}$$

where $\beta(T) = \alpha'/c_p$.

In the ESP integration procedure, (8.21) is used to calculate dP_c/dt, taking $\beta = (\gamma - 1)/\gamma$, where $\gamma = c_p/c_v$, for each zone. Then (8.17, 8.18) are used to advance the total enthalpies in the two zones. The zone mass balances are advanced using (8.13, 8.14). The new masses and total enthalpies determine the zone temperatures. The unburned volume is then calculated from the volume constraint (8.15), written as

$$\frac{V_u}{V_c} = \frac{(MRT)_u}{(MRT)_u + (MRT)_b} \tag{8.22}$$

and the burned volume is just the remainder. Finally, the pressure is calculated from $PV = MRT$ for whichever zone is larger.

The procedure outlined above requires values for the burn rate $\dot{M}_b$ and the heat transfer rates $\dot{Q}_u$ and $\dot{Q}_b$. The heat transfer rates are determined by

$$\dot{Q}_u = St_u V_{tu} \rho_u c_{pu} A_u (T_u - T_w) \tag{8.23}$$

and

$$\dot{Q}_b = St_b V_{tb} \rho_b c_{pb} A_b (T_b - T_w) \tag{8.24}$$

where St_u and St_b are used-specified Stanton numbers for the two zones based on the turbulence velocities V_{tu} and V_{tb} in each zone, computed from the turbulent kinetic energy in each zone. The heat transfer areas A_u and A_b are determined from a user-input flame geometry table and the total surface area. These factors are discussed in more detail below.

The burn rate is given by

$$\dot{M}_b = A_f \rho_u V_f \tag{8.25}$$

where A_f is the projected area of the flame front (i.e., the area of a smooth, spherical surface faired through the instantaneous location of the wrinkled flame), ρ_u is the density of the unburned gas approaching the front, and V_f is the velocity of the front relative to the unburned gas. V_f is in turn given by

$$V_f = V_L + C_f V_{tu} \tag{8.26}$$

where V_L is the specified laminar flame speed and C_f is a specified coefficient, approximately unity. Thus, turbulence increases the burn rate, as in real engines. The turbulence model, described below, tracks the turbulent kinetic energy in each zone. It includes the important effect of turbulence enhancement in the unburned gas caused by the rapid compression, which in turn tends to make the flame burn faster.

The *flame geometry table* provided by the user is used to calculate the heat transfer and flame areas defined above. The table gives the ratio of each of these areas to the bore area as a function of the fraction of the volume occupied by burned gas, from 0 to 1. The table is generated in a separate geometrical calculation in which the user assumes (for example) a spherical flame propagating from the spark plug to determine the pertinent areas for various flame positions. In the present version of ESP, this table is constructed for one piston position (typically TC). The heat transfer area ratios are applied to the total surface area exposed to the gas at the current crank angle, and the flame area inferred from the table is multiplied by the ratio of the cylinder volume to the clearance volume as a means for first-order consideration of the increased flame area that is expected as the cylinder volume expands.

This approach to flame geometry allows a user to make preliminary studies of the effects of spark plug placement and combustion chamber geometry on engine performance. If used with due regard for the great simplicities in the model, the approach can help a designer move in the right general direction.

The user specifies a combustion efficiency (fraction of reactants in the cylinder converted to products during the burn). The burn analysis ends when that has been achieved, or when the exhaust valve opens. The burned and any residual unburned gases are assumed to mix instantaneously and adiabatically at the end of the burn stage. The mixed temperature and pressure are computed from the total internal energy and volume. The mixed turbulence kinetic energy is determined by mass-weighting.

8.3.5 Analysis of the Expansion Stage

The expansion stage is handled exactly the same as the compression stage. However, the user may specify a different set of model constants in the heat transfer and turbulence models for this stage.

8.3.6 Analysis of the Gas Exchange Stage

During gas exchange, the energy balance on the expanding and contracting control volume consisting of the in-cylinder gases is

$$\begin{pmatrix} h_i \dot{M}_i - \dot{Q}_{iv} \\ -h_c \dot{M}_{ib} \end{pmatrix} = \frac{dU_c}{dt} + \dot{W}_p + \dot{Q}_c + \begin{pmatrix} h_c \dot{M}_e \\ -h_e \dot{M}_{eb} + \dot{Q}_{ev} \end{pmatrix} \tag{8.27}$$

This is shown symbolically in Figure 8.5. The left-hand side represents terms associated with the intake valve, the upper terms being used if there is inflow and the lower if there is backflow through the intake valve. Similarly, the last terms are the exhaust valve terms, the upper being used when there is outflow and the lower with exhaust backflow. Here h_i is the specific stagnation enthalpy of the incoming gas and $\dot{M}_i$ is the inlet valve mass inflow rate, h_c is the specific enthalpy of the in-cylinder gas, $\dot{M}_{ib}$ is the intake valve mass backflow rate, $\dot{Q}_{iv}$ is the heat transfer rate to the intake valve area, $\dot{M}_e$ is the exhaust valve mass outflow rate, h_e is the specific stagnation enthalpy of the mixed exhaust gases, $\dot{M}_{eb}$ is the mass backflow rate through the exhaust valve, and $\dot{Q}_{ev}$ is the heat transfer rate to the exhaust valve area. $\dot{W}_p$ and $\dot{Q}_c$ are as defined and computed above.

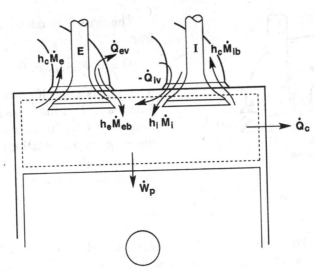

Figure 8.5. Cartoon of the energy balance on the in-cylinder gases, with the various terms in Equation 8.27 indicated.

If there is intake backflow or if there is any backflow gas in the intake manifold, then an energy balance on the backflow zone is also used:

$$\left(\begin{matrix} h_c \dot{M}_{ib} - \dot{Q}_{iv} \\ -h_i \dot{M}_i \end{matrix} \right) = \frac{dH_{ib}}{dt} \tag{8.28}$$

This is shown symbolically in Figure 8.6. The total enthalpy H_{ib} of the backflow gas arises when one accounts for the total internal energy change in the backflow zone plus the work done in pushing back the fresh charge at the constant intake manifold pressure P_i. The upper terms on the left are used with backflow and the lower terms when backflow gas is entering the cylinder.

An energy balance is also made on the exhaust gas zone (mixed over each cycle):

$$\left(\begin{matrix} h_c \dot{M}_e - \dot{Q}_{ev} \\ -h_e \dot{M}_{eb} \end{matrix} \right) = \frac{dH_e}{dt} \tag{8.29}$$

This is shown symbolically in Figure 8.7. The total enthalpy H_e of the exhausted gas arises when one accounts for the total internal energy change in the backflow zone plus the work done in pushing away the exhaust gas from the last cycle at the constant exhaust manifold pressure P_e. The upper terms on the left are used with outflow and the lower terms with exhaust gas backflow.

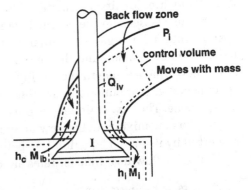

Figure 8.6. Cartoon of the energy balance of the backflow zone, with the terms of Equation 8.28 indicated.

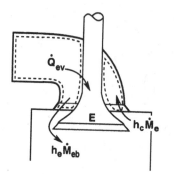

Figure 8.7. Cartoon of the energy balance on the exhaust gas zone, with the terms of Equation 8.29 indicated.

The mass flow rates are determined from

$$\dot{M} = C_d F(A, \rho, P, P_b, \gamma) \qquad (8.30)$$

where C_d is a specified discharge coefficient and F is the mass flow rate for isentropic flow in a contraction from the upstream (stagnation) state to the current valve flow area, as calculated for the contraction area A, upstream density ρ, upstream pressure P, downstream (back) pressure P_b, and gas specific heat ratio $\gamma = c_p/c_v$. Flow area is calculated as in Equation 2.1. The flow routine provides for choked flow and has special provisions for accuracy at very low flow rates.

Mass balances are also required. For the cylinder zone,

$$\begin{pmatrix} \dot{M}_i \\ -\dot{M}_{ib} \end{pmatrix} = \frac{dM_c}{dt} + \begin{pmatrix} \dot{M}_e \\ -\dot{M}_{eb} \end{pmatrix} \qquad (8.31)$$

For the intake backflow zone,

$$0 = \frac{dM_{ib}}{dt} + \begin{pmatrix} \dot{M}_i \\ -\dot{M}_i b \end{pmatrix} \qquad (8.32)$$

For the exhaust gas zone,

$$\begin{pmatrix} \dot{M}_e \\ -\dot{M}_{eb} \end{pmatrix} = \frac{dM_e}{dt} \qquad (8.33)$$

In these mass balances, the upper terms are used for normal valve flow and the lower terms for backflow.

To advance to the next time step, the pertinent flow rates and associated isentropic velocities are first calculated. The mass balances are then used to advance the various masses and the energy balances to advance U, H_e, and H_i. Then the specific internal energies (or enthalpies) of these zones are calculated by dividing by the new masses, and the amounts of reactants and products in each zone at the new time are determined using the mass exchange information. The new specific energies or enthalpies are then used to determine the new temperatures of each zone. The new cylinder-gas pressure is then calculated from the Pv product for the new mixture at the new temperature. The turbulence kinetic energy is advanced by its model equation (see below).

The valve flow heat transfer rates are determined from equations of the form

$$\dot{Q}_v = St_v V_v \rho c_p A_v (T - T_w) \qquad (8.34)$$

Here St_v is a user-specified Stanton number based on the velocity V_v for isentropic flow through the valve; ρ, c_p, and T are the upstream density, specific heat, and temperature; and A_v is an effective area for this heat transfer, specified by the user. These rates are clipped to keep the backflow and exhaust gas zones from dropping below the specified wall temperature; this provides a quick solution to a minor numerical problem that sometimes occurs when the zone mass is small, and does not significantly affect the cycle performance predictions.

8.3.7 Turbulence Model

The turbulence model, which is a unique feature of ESP, is based on equations that describe the evolution of turbulence in a fluid where the density is a function only of time. For details see [51]. The turbulence kinetic energy per unit mass k (averaged over the volume) is described by

$$\frac{dk}{dt} = P - D + \frac{1}{M_c}\left(I - E - k\frac{dM_c}{dt}\right) \tag{8.35}$$

The meaning and modeling of the terms on the right will now be discussed.

P is the turbulence energy production rate per unit mass, which turbulence theory shows is the product of the strain rate of the mean motion with the Reynolds stresses of the turbulence. The production model must include both the effects of strain in the shear flows driven by the mean piston motion and the effects of turbulence enhancement due to compression. V_c/A_w, the ratio of cylinder volume to surface area, represents a length scale for the mean strain rate, so the mean strain rate should vary as $|V_p| A_w \backslash V_c$, where V_p is the piston speed. The turbulent stresses have the dimension of *velocity*^{2}. Model tests suggested that V_p^2 was a better scale than $k = V_t^2 \backslash 2$ for the stresses, and so P is modeled by

$$P = F_p \frac{A_w}{V_c}|V_p|^3 - \frac{2}{3}k\frac{1}{V_c}\frac{dV_c}{dt} \tag{8.36}$$

The first term accounts for turbulence production due to strain in the shear flow on the walls and the second the effects of compression. The coefficient F_p is user-selected and is one of several model parameters that can be set by reference to experiments. See the sample ESP setup files for typical values. The 2/3 coefficient in the second term is given exactly by *rapid distortion theory* [51], with which this model is therefore consistent.

D is the turbulence energy dissipation rate per unit mass, which turbulence theory indicates should vary as V_t^3/L, where L is the scale of the largest turbulent eddies. Turbulence in an engine has a scale comparable with that of the cylinder, so the dissipation model is

$$D = F_d \frac{kV_t}{V_c^{1/3}} \tag{8.37}$$

where F_d is a user-selected model constant.

Equation (8.35) is obtained by integrating the partial differential equation for the turbulent kinetic energy over the cylinder volume, denoting the average kinetic energy per unit mass by k. This gives rise to terms associated with flow through the valves, denoted by I and E for the intake and exhaust valves. Flow visualization experiments at Stanford suggest that most of the kinetic energy of the mean flow entering through the valves is converted to turbulence energy in the cylinder. Therefore, the model for the intake term is

$$I = \begin{pmatrix} F_i \dot{M}_i V_i^2/2 \\ -\dot{M}_{ib}k \end{pmatrix} \tag{8.38}$$

The upper term on the right is used when there is inflow through the intake valve and the lower with backflow. F_i is a factor, less than unity, set by the user. Similarly, for the exhaust term

$$E = \begin{pmatrix} \dot{M}_e k \\ -F_e \dot{M}_{eb} V_{eb}^2/2 \end{pmatrix} \tag{8.39}$$

The upper term on the right is used when there is discharge through the exhaust valve and the lower with backflow. F_e is another user-set factor.

During the burn stage the turbulence energies of the burned and unburned gases are tracked separately using the above equations with just the P and D terms. The compression-production term for the unburned gas is based on its rate of compression and is important because it increases the turbulence level hence the burn rate. The compression-production term for the burned gas is calculated on the basis of the overall volume change-rate for the cylinder. Since at the end of the burn the cylinder contains essentially all burned gas, the overall effect of the compression-production term during burn is just that caused by the piston motion. These assumptions put some elements of reality into the turbulence model without assuming more detail that is really justified.

8.4
ESP MANIFOLD ANALYSIS

8.4.1 Overview

MOTIVATION
This analysis grew out of Lumley's suggestion that Reynolds modify his Engine Simulation Program (ESP) to consider intake and exhaust manifold dynamics. We want to have a good representation of the unsteady gasdynamics in the manifolds, but we do not want the complexity of solving one-dimensional time-dependent PDEs. Therefore, we need a good way to treat the unsteady manifold flows using ordinary differential equations for the properties at nodes in the network as functions of time.

LUMPED PARAMETER APPROACH
One possibility would be to use a lumped parameter approach, in which the capacitance for mass accumulation is lumped at the branch points or plenum

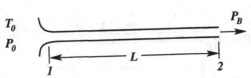

Figure 8.8. Duct fed by a contraction from a plenum.

chambers and the resistance to flow is placed in the ducts. Such a model can give a good description of Helmholtz resonators, for example that provided by the air filter plenum. However, because it ignores compressibility in the ducts, the mass flow rate out of the intake runner at any moment must be the same as the mass flow rate in, whereas in reality the inflow lags the outflow because of accumulation in the duct. The lumped parameter approach cannot capture this important effect. Because of this, Reynolds developed a new approach outlined here.

OUTLINE OF THE MODEL

Experiments and analysis with and without friction show that friction in engine manifold flows is relatively unimportant compared to the unsteady compressible flow dynamics, and that a quite accurate manifold prediction can be obtained by neglecting friction (See for example [88], Volume 2, Figure 24.27). In our model friction is treated approximately, as a perturbation to the dominant unsteady gasdynamics. The basic flow element is a length of duct fed from a plenum by a contraction, Figure 8.8. The contraction flow is treated as quasi-steady and isentropic. Analysis based on the method of characteristics is then employed to analyze the unsteady adiabatic one-dimensional compressible ideal-gas flow in the constant-area ducts, with some simplifying approximations that allow the problem to be solved using retarded ordinary differential equations (ODEs involving both the current and retarded times) at the branch points in the manifolds, based on analytical solutions of the PDEs used in the method of characteristics. The text provides some background on the ideas of characteristics followed by an outline of the manifold model.

The model we will outline here is thus a combination of the organ pipe model and the Helmholtz resonator model described in Chapter 2.

8.4.2 Unsteady One-Dimensional Compressible Flow

This subsection amplifies the discussion of the organ pipe model in Section 2.12.3 of Chapter 2.

ACOUSTIC WAVES TO THE RIGHT

Figure 8.9 shows the effect of a small increase in stagnation pressure δP_0 when initially $P_0 = P_B$ (no flow). The increase in stagnation pressure initiates a compression wave running to the right at the speed of sound c relative to the fluid into which it is travelling, which in this case is motionless ($V_{\text{fluid}} = 0$). The fluid ahead of the wave is at rest, and the fluid behind the wave moves to the right with

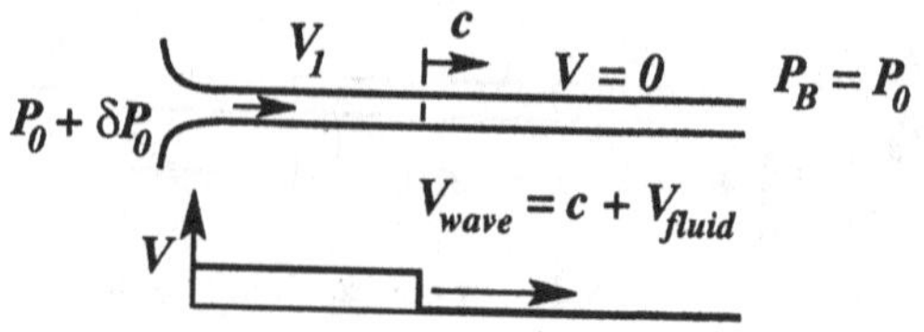

Figure 8.9. Acoustic wave moving right as a result of stagnation pressure increase.

a small mass velocity V_1. The fluid behind the wave is compressed to a slightly higher density (which is why it is called a compression wave). If nothing else changes, the compression wave will reach the end of the duct after a time interval of L/c.

ACOUSTIC WAVES TO THE LEFT
Figure 8.10 shows the effect of a small decrease in the back pressure δP_d when initially $P_B = P_0$ (no flow). The decrease in back pressure initiates an expansion wave running to the left at the speed of sound c relative to the fluid into which it is travelling, which in this case is motionless ($V_{fluid} = 0$). The fluid ahead of the wave is at rest, and the fluid behind the wave moves to the right with a small velocity V_1. The fluid behind the wave is expanded to a slightly lower density (which is why it is called an expansion wave). If nothing else changes, the compression wave will reach the entrance to the duct after a time interval of L/c.

SEQUENCE OF ACOUSTIC WAVES TO THE RIGHT
Figure 8.11 shows the effect of a series of small increases in the stagnation pressure. Each increase generates a new right-running compression wave. Each compression wave raises the temperature. Therefore, the sound speed is greater behind the first wave, and so the second wave runs faster than the first. Moreover, the flow velocity is greater after each compression, and so each subsequent wave is convected faster by the fluid than the waves ahead of it. The net effect is that the waves behind will catch the waves in front; this is how shock waves form.

SEQUENCE OF ACOUSTIC WAVES TO THE LEFT
Figure 8.12 shows the effect of a series of small decreases in the back pressure. Each decrease generates a new left-running expansion wave. Each expansion wave

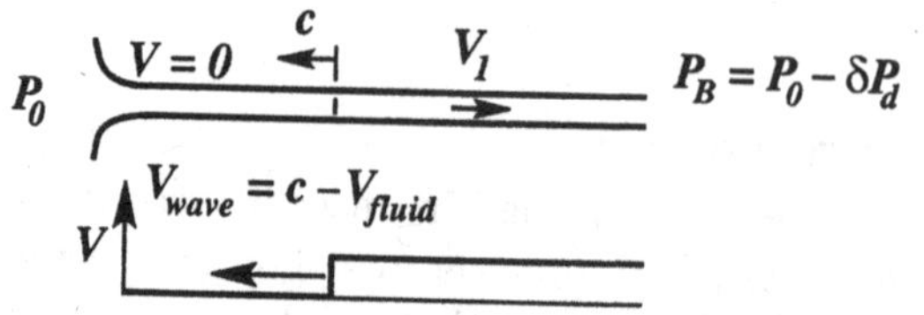

Figure 8.10. Acoustic wave moving left as a result of back pressure decrease.

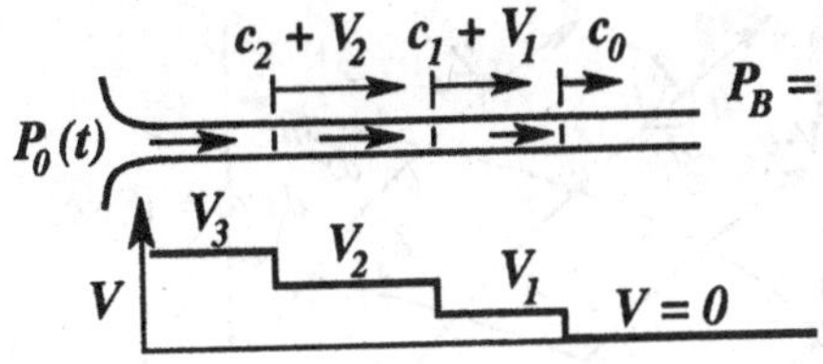

$$c_2 + V_2 \quad c_1 + V_1 \quad c_0$$

$$P_B = constant$$

$$P_0(t)$$

$$V_3 \quad V_2 \quad V_1$$

$$V \qquad V = 0$$

Figure 8.11. Sequence of right-running acoustic waves arising from a series of stagnation pressure increases.

lowers the temperature. Therefore, the sound speed is less behind the first wave, and so the second wave runs slower than the first. Moreover, the flow velocity is greater after each compression, and so each subsequent wave is retarded by a greater headwind than the waves ahead of it. The net effect is that the waves will spread out as they travel upstream against the flow.

DOWNSTREAM ACOUSTIC CHARACTERISTICS
The path of a right-running acoustic wave is defined by

$$\frac{dx}{dt} = V + c \tag{8.40}$$

where c and V are the sound speed and velocity of the flow into which the wave is moving. These path lines are called the *downstream acoustic characteristics*. The discussion above shows that information from the upstream is carried downstream along these characteristics. This information is denoted by Z^+ in Figure 8.13.

UPSTREAM ACOUSTIC CHARACTERISTICS
The path of a left-running acoustic wave is defined by

$$\frac{dx}{dt} = V - c \tag{8.41}$$

where c and V are the sound speed and velocity of the flow into which the wave is moving. These path lines are called the *upstream acoustic characteristics*. The discussion above shows that information from downstream is carried upstream along these characteristics. This information we denote by Z^- in Figure 8.13.

CONVECTION CHARACTERISTIC
In addition to the two families of acoustic characteristics, there is a third family of characteristics associated with downstream convection of information by the flow. The path of the convection characteristic is given by

$$\frac{dx}{dt} = V \tag{8.42}$$

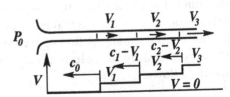

Figure 8.12. Sequence of left-running acoustic waves arising from a series of back pressure decreases.

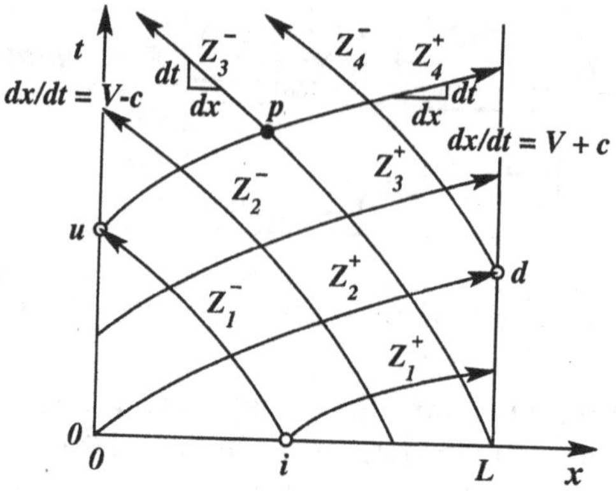

Figure 8.13. Acoustic characteristics in the x-t plane.

MACH LINES AND PATH LINES

The acoustic characteristics are sometimes called *Mach lines*. The information they carry travels through the fluid at the speed of sound. The convection characteristic is sometimes called the *path line* because it follows the path of a fixed fluid particle. The information it carries travels downstream with the fluid.

8.4.3 The Method of Characteristics

CONCEPT OF CHARACTERISTICS

An exact calculation of this unsteady flow would involve the solution of the partial differential equations governing unsteady one-dimensional gasdynamics. This calculation could be done using the *method of characteristics* (See for example [88], Volume 2, Art. 24.7.). In a one-dimensional time-dependent PDE problem, a characteristic is a line in x-t space along which the basic set of first order PDEs can be combined to produce a single first-order quasi-ordinary differential equation containing derivatives only along the characteristic. These are precisely the acoustic and convection characteristics described above.

FRICTIONLESS ADIABATIC FLOW

For the case of frictionless adiabatic flow of an ideal gas through a constant area duct, the PDEs show that some property of the flow is constant on each characteristic. This feature makes this particular flow model very easy to handle. We outline the equations that apply for this case, after which we give the modifications necessary to include friction, which is of minor importance in engine manifold flows.

CONSTANT ON THE CONVECTION CHARACTERISTIC

Because for frictionless adiabatic flow the entropy of each fluid particle is constant the entropy is constant on a convection characteristic. Changes in the entropy are carried down the duct by the mass and move along the convection characteristics. For computation it is convenient to define

$$Y = \frac{P}{\rho^k} \tag{8.43}$$

because the entropy s can be taken as simply

$$s = c_v \ln Y \tag{8.44}$$

For frictionless adiabatic flow, Y is constant on a convection characteristic.

CONSTANTS ON THE ACOUSTIC CHARACTERISTICS

For frictionless adiabatic flow, the quasi-ODEs on the acoustic characteristics show that the quantity

$$Z^+ = \frac{2}{k-1}c + V \text{ (downstream characteristics)} \tag{8.45}$$

is constant on a downstream acoustic characteristic (k is the specific heat ratio c_p/c_v) and the quantity

$$Z^- = \frac{2}{k-1}c - V \text{ (upstream characteristics)} \tag{8.46}$$

is constant on an upstream characteristic. Thus, the values of Z^+ and Z^- at the point (x, t) where these two characteristics intersect determine V and c. For example, in Figure 8.13 at the point p

$$\frac{2}{k-1}c + V = Z_4^+ \qquad \frac{2}{k-1}c - V = Z_3^-, \tag{8.47}$$

hence at this point

$$c = (Z_4^+ + Z_3^-)\frac{k-1}{4} \qquad V = \frac{1}{2}(Z_4^+ - Z_3^-) \tag{8.48}$$

Using the ideal gas equation of state, the value of the entropy along the convection characteristic through this point, together with the temperature T determined from $c = \sqrt{kRT}$, determines the pressure P and density ρ.

DETERMINING Z^+ AND Z^- FROM INITIAL CONDITIONS

The values of Z^+ and Z^- are determined by the boundary or initial conditions at the point on the (x, t) boundary where the characteristics originate. For example, at point i of Figure 8.13 the initial values of V_i and c_i determine Z_1^+ and Z_1^- using Equations 8.43, 8.44.

DETERMINING Z^- AT A DOWNSTREAM BOUNDARY

The value of Z^- at a downstream boundary is determined by the value of Z^+ coming to that point and a condition at the boundary. For example, if V is known at point d of Figure 8.13, then

$$c_d = \frac{k-1}{2}(Z_2^+ - V_d) \qquad Z_4^- = \frac{2}{k-1}c_d - V_d \tag{8.49}$$

DETERMINING Z^+ AT AN UPSTREAM BOUNDARY

The value of Z^+ at an upstream boundary is determined by the value of Z^- coming to that point and a condition at the boundary. For example, if V is known at point u of Figure 8.13, then

$$c_u = \frac{k-1}{2}(Z_1^- + V_u) \qquad Z_4^+ = \frac{2}{k-1}c_u - V_u \tag{8.50}$$

EQUATIONS WITH FRICTION AND ENTROPY CHANGE

For one-dimensional unsteady adiabatic flow of an ideal gas in a constant area duct with friction, the equation along a convection characteristic is

$$\left.\frac{d\ln Y}{dt}\right|_{\text{conv.}} = k(k-1)\frac{1}{2}f\frac{b}{A}M^2 V \tag{8.51}$$

Here f is the friction factor defined in terms of the wall shear stress $\tau_w = \frac{1}{2}\rho V^2 f$, b is the wetted perimeter, A is the flow area, and $M = V/c$ is the local Mach number. The equations along the two types ($\pm$) of acoustic characteristics may be written as[2]

$$\left.\frac{dZ^\pm}{dt}\right|_{\pm\text{acou.}} = \frac{c}{k(k-1)}\left.\frac{d\ln Y}{dt}\right|_{\pm\text{acou.}} + \frac{1}{2}f\frac{b}{A}[(k-1)|M|V^2 \mp V|V|] \tag{8.52}$$

Note that if the flow is frictionless ($f = 0$) so that $Y = P/\rho^k$ is constant, $Z^\pm$ is constant along the acoustic characteristic having $dx/dt = V \pm c$, as described above. The $\ln Y$ derivatives are not the same in Equation 8.51 and Equation 8.52 because they are derivatives along different characteristics. We see that the effect of friction is to increase the entropy (or equivalently $Y = P/\rho^k$) along the convection characteristics and modify the Z values on the acoustic characteristics. These changes are small for typical engine manifold flows, and so friction effects can be included as a modest perturbation on the frictionless analysis outlined above. The first term in (8.52) includes the effect of entropy gradients along the characteristic. These are very important in the exhaust manifold, and ESP models this term in a way that is exact for sharp entropy discontinuities.

---

[2] In the first printings of [88], Volume 2, a factor of k is missing before the Mach number in the friction term of Shapiro's (24.52). We have expressed the entropy partial derivative with respect to x at fixed t appearing in his (24.52) in terms of the entropy gradients along the acoustic characteristic and along the local convection characteristic.

EXACT CHARACTERISTICS CALCULATION

An exact treatment of this flow by the method of characteristics involves numerical integration of the equations on the characteristics (Equations 8.51, 8.52) simultaneously with the integration of the characteristic path equations (Equations 8.40, 8.41 and 8.42) using the local Z^+, Z^-, and state (Y) values determined by the initial and boundary conditions. This requires storage of information as functions of both space and time. Since this is more computationally intensive than desired for ESP, some approximations are made that eliminate the need for storing information except at the ends of each duct. These approximations are described in the following sections.

APPROXIMATE ACOUSTIC CHARACTERISTICS

We denote the speeds for the right- and left-running acoustic characteristics by

$$V^+ = c + V$$
$$V^- = c - V \tag{8.53}$$

Both V and c change on the characteristics, but in opposite directions. As the flow accelerates, the temperature (and hence c) drops, and vice versa. For low Mach numbers the main contributor to the wavespeed is c, which is a weak function of the absolute temperature. Because the absolute temperature does not change very much, the acoustic characteristics are very nearly straight. However, ESP does *not* assume that they are straight. Instead it uses a time-delay model to determine Z^+ and Z^- in terms of the values at the boundaries at earlier times, as shown in Figure 8.14.

ACOUSTIC TIME DELAY MODEL

ESP determines the acoustic time delays using the model equation

$$\frac{dt_{d\pm}}{dt} = 1 - \frac{1}{\tau_\pm} t_{d\pm} \tag{8.54}$$

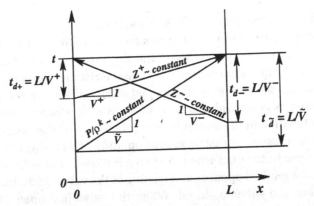

Figure 8.14. Approximate acoustic characteristics.

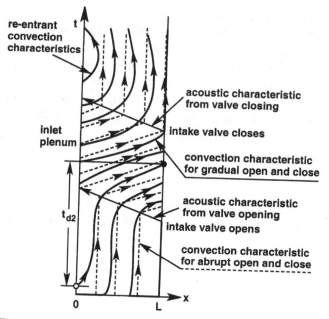

re-entrant
convection
characteristics

acoustic characteristic
from valve closing

inlet
plenum

intake valve closes

convection characteristic
for gradual open and close

acoustic characteristic
from valve opening

intake valve opens

convection characteristic
for abrupt open and close

Figure 8.15. Convection characteristics.

where

$$\tau_{\pm} = \frac{L}{2}\left(\frac{1}{c_1 \pm V_1} + \frac{1}{c_2 \pm V_2}\right) \tag{8.55}$$

This model has several important features. It drives $t_{d\pm}$ to τ on the time scale τ for crossing of an acoustic characteristic. If $V_1 = V_2 = 0$ and $c_1 = c_2 = c$ then the time delay is quickly driven to L/c. If $c - V$ is small at one end of the duct, the delay time is long and controlled by $c - V$ at that end. Since $dt_{d\pm}/dt < 1$, a characteristic that arrives at one end after another characteristic cannot have been launched before the launching of the other characteristic. ESP advances the time delays using Equation 8.54, and then determines the values of $Z^{\pm}$ at the launch times from the recorded history of V and c at the ends of the duct, with modeled corrections for the friction integrals and entropy change.

NATURE OF THE CONVECTION CHARACTERISTICS

For frictionless flow of fluid that enters the duct with a constant entropy, it is not necessary to assume anything about the shape of the convection characteristics because the entropy is the same on all characteristics. However, if the entropy changes due to changes in the inlet entropy (or friction), some assumption about these characteristics is needed. Figure 8.15 shows the general nature of the convection characteristics in the runner to the intake valve. The dashed lines show the characteristics assuming instantaneous opening and closing of the valve, neglecting oscillations in the duct when the valve is closed. The solid lines show the characteristics for gradual opening and closing of the valve, including oscillations in the duct when the valve is closed. When the valve first opens, the fluid that initially leaves the duct at L had entered during the previous intake cycle. After the acoustic characteristic from the valve opening has reached the inlet ($x = 0$) new gas begins to enter the runner, and initially this gas will travel through the duct and exit at $x = L$. When the intake valve closes, the gas in the runner must remain there, perhaps oscillating back and forth due to compressibility, until the valve opens for the next engine cycle.

CONVECTION CHARACTERISTIC TRACKING

Because of the oscillations in the ducts, there are some convection characteristics that enter one end and leave through the same end. Figure 8.15 includes one of these characteristics, which we call re-entrant to distinguish them from characteristics that cross the duct (crossing characteristics). Thus the convection characteristics can be very complicated. In order to track the entropy of the mass moving about in the ducts, ESP calculates and records the total mass that has passed the inlet and exit of each duct since the calculation began and the entropy (Y) value of the fluid when it entered. The amount of mass passed at the end where mass is leaving is then used to determine when the mass entered and from which end, and the initial entropy of that mass is then looked up in the history record of mass entering the duct at that end.

FRICTION PERTURBATIONS FOR THE CONVECTION CHARACTERISTICS

We use the subscript 1 to denote the time and place at which a characteristic is launched in the x-t diagram, and the subscript 2 to denote the later time and place where it is received. Then, integrating Equation 8.51 along the characteristic, we have

$$\ln Y_2 - \ln Y_1 = \int_1^2 k(k-1)\frac{1}{2}f\frac{b}{A}M^2\,V dt \tag{8.56}$$

Since f is a weak function of velocity, we take it as constant. Note that on a convection characteristic $V\,dt = dx$. ESP approximates the integral and uses

$$\frac{Y_2}{Y_1} = \exp\left(k(k-1)\frac{1}{2}f\frac{b}{A}\frac{V_1^2 + V_2^2}{2\bar{c}^2}L\right) \tag{8.57}$$

where $\bar{c} = (c_1 + c_2)/2$. For re-entrant convection characteristics ESP assumes that the characteristic path is a symmetric parabola with equal but opposite slopes dx/dt at the entry and exit times, which leads to the approximation

$$\frac{Y_2}{Y_1} = \exp\left(k(k-1)\frac{1}{2}f\frac{b}{A}\frac{|V_2 - V_1|^3(t_2 - t_1)}{32\bar{c}^2}\right) \tag{8.58}$$

FRICTION CORRECTIONS FOR THE ACOUSTIC CHARACTERISTICS

Similar approximations are made for the acoustic characteristics. However, on these characteristics we cannot use $V\,dt = dx$. Integrating Equation 8.52,

$$Z_2^{\pm} - Z_1^{\pm} = \int_1^2 \frac{c}{k(k-1)}d\ln Y + \int_1^2 \frac{1}{2}f\frac{b}{A}[(k-1)\,|M|\,V^2 \mp V|V|]\,dt \tag{8.59}$$

On the acoustic characteristics, $dx/dt = V \pm c$, so $dx = \pm dt/(c \pm V)$. Then trapezoidal integration leads to the ESP approximations

$$Z_2^{\pm} - Z_1^{\pm}$$

$$= \frac{2}{k-1}[c(P_1, Y_2) - c(P_1, Y_1)] + \frac{1}{2}f\frac{bL}{2A}\left[\pm(k-1)\left(\frac{|V_1|V_1^2}{c_1(c_1 \pm V_1)} + \frac{|V_2|V_2^2}{c_2(c_2 \pm V_2)}\right)\right.$$
$$\left. - \left(\frac{V_1|V_1|}{c_1 \pm V_1} + \frac{V_2|V_2|}{c_2 \pm V_2}\right)\right]$$

$$(8.60)$$

where $c(P, Y)$ is the speed of sound evaluated as a function of P and Y.

FRICTION IS NOT VERY IMPORTANT BUT CRUCIAL

For a round duct of diameter D, the factor fbL/A is $4fL/D$; estimating $4f = 0.01$ and $L/D = 10$, for $M = 0.7$ the exponential factor in Equation 8.57 increases Y by less than 2%. This shows that friction is a small perturbation for which the approximations described above should suffice. However, friction is still crucial because it limits the build-up of resonances. Hence it has to be included.

GAS PROPERTIES MODEL

The manifold analysis in each manifold is based on ideal gas with constant specific heats.

VALVE FLOW AND ENTRANCE AND EXIT BLOCKAGE

At the heart of ESP is a subroutine that calculates the flow to (or from) a plenum through a flow restriction into (or out of) a duct. This is used for the intake and exhaust valves, where the plenum is the cylinder, for the entrance and exit to ambient, and at the junction, where the junction is treated as the plenum. The state in the plenum and the Y and Z values arriving on the characteristics determine the flow through the restriction and the state in the duct. This end-flow subroutine allows the flow to be choked at the point of minimum area or in the duct should conditions of that serverity arise (as a result of strong entropy gradients).

For flow into the plenum, the flow is assumed to be isentropic from the duct to the point of effective minimum area. The effective minimum area is the actual minimum flow area times the user-specified flow coefficient. The minimum area for a valve is the instantaneous valve flow area. The minimum area at other duct ends is set by the user-specified duct blockage fraction, which can be used to account for flow losses or for a throttle. The pressure at the point of minimum area is assumed to be equal to the plenum pressure unless the flow is choked, in which case sonic flow is assumed.

For entry into the duct, the flow from the plenum to the point of effective minimum area is assumed to be isentropic, and the discharge jet pressure at the minimum area point is assumed to be the back pressure. The pressure recovery after the contraction is calculated using momentum, mass, and energy balances.

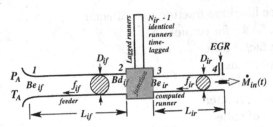

Figure 8.16. Inlet Manifold Model.

The analysis can handle a sharp entropy discontinuity as occurs in the exhaust manifold when very hot combustion products first come blasting out.

8.4.4 Inlet Manifold Model

Figure 8.16 shows the inlet manifold model and parameters.

Be_{if}	entrance blockage fraction for feeder
Bd_{if}	discharge blockage fraction for feeder
Be_{ir}	entrance blockage fraction for runner
f_{if}	friction factor for feeder
f_{ir}	friction factor for runner
L_{if}	length of feeder
L_{ir}	length of runner
D_{if}	diameter of feeder
D_{ir}	diameter of runner
N_{ir}	number of runners fed by the feeder

8.4.5 Exhaust Manifold Model

Figure 8.17 shows the exhaust manifold model and parameters.

Bd_{er}	discharge blockage fraction for runner
Be_{ec}	entrance blockage fraction for collector

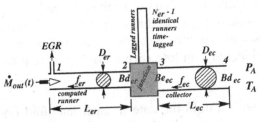

Figure 8.17. Exhaust Manifold Model.

Bd_{ec} entrance blockage fraction for collector
f_{er} friction factor for runner
f_{ec} friction factor for collector
L_{er} length of runner
L_{ec} length of collector
D_{er} diameter of runner
D_{ec} diameter of collector
N_{er} number of runners feeding the collector

8.4.6 ESP Calculations

COMPUTED RUNNER

ESP calculates only one runner and cylinder. The flow to or from the other runners is determined from that for the computed runner at an appropriate earlier time in the engine cycle. The duct calculations are made using the approximate method of characteristics outlined above. The manifold and cylinder calculations are coupled at the valves by the end-flow subroutine. The ordinary differential equations in both the cylinder and manifolds are integrated numerically using a second-order Runge-Kutta method.

ENTRY

The entry to the intake manifold is assumed to be an isentropic contraction, followed by the entrance loss and then the feeder. The calculation of intake manifold state 1 requires Z^- carried to point 1 by the $-$ acoustic characteristic originating at point 2, determined using the acoustic time delay as described above.

JUNCTION

The junction stagnation state is determined by quasi-steady mass and energy balances on the junction, which are solved iteratively with the end-flow subroutine so that the duct end states are fully consistent with the arriving characteristic data (Y, Z) and the junction state. For multi-cylinder manifolds, the runner state (Y, Z) at the junction in the computed runner is phase-averaged, and the flow to/from the non-computed runners are computed using the appropriate phase-average Y and Z by the end-flow subroutine.

PERIODICITY

After a few hundred engine cycles a periodic solution should be obtained. However, for multi-cylinder configurations the calculation during the transient is really not right because only one of the runners is actually computed. In order to reach the periodic solution (which *is* correct within the modeling approximations) as quickly as possible, ESP employs several tricks that will now be outlined.

TRICK ONE

ESP starts out with a uniform state in each manifold duct, which sets the initial mass in the duct. But as the flow develops this mass will change, and the change must be considered in order to get the right exit entropy in the convection characteristic analysis. After each engine cycle, ESP determines the excess mass that entered or left the duct, and then flushes the mass appropriately. Eventually the mass in the duct become stable and the flushing ceases.

TRICK TWO

After each time advance, the solution for the end states of the feeder is replicated with appropriate phase shift to render the feeder solution periodic at the frequency appropriate for the number of cylinders on the manifold. This forces the proper periodicity in the feeder, but means that the solution is not time-accurate until it has converged to a solution that repeats itself on each engine cycle. The manifold pressure and velocity plots help assess the convergence. The convergence is also observable in the one-line display of the mass that passed each manifold point in the previous cycle.

TRICK THREE

Given the discussion above, you might think that ESP would be time-accurate for a single cylinder case, but even that is not true because ESP only remembers data in the manifold for one engine cycle of 720 degrees. If the convection analysis indicates that the gas emerging from one end has been in the duct for more than one engine cycle (often the case), ESP infers the amount of mass that passed each end more than 720 degrees earlier using the data for the 720 remembered degrees, assuming periodicity. This means that for the first several computational cycles the manifold states can be very wrong. In the case of the exhaust manifold, early cycles can give the flow past one or more points with the wrong sign, but don't let this bother you when you see it on the ESP manifold convergence monitor. In due course (two or three more cycles) ESP will sort this out and the manifold flow will approach a converged repetitive pattern.

TRICK FOUR

If you run ESP with a single cylinder on the manifold, you will see that all four flows on the convergence monitor are the same. This should give you some confidence that the junction routine is properly conserving mass. So does the manifold duct analysis, but not exactly because of the numerical and physical approximations made. Therefore, the convection characteristic analysis will slowly drift, and this is fixed by a third trick. After each cycle, the mass unbalance for each duct is computed, and the amount (very small) that the Z values would have to change to make an exact mass balance is determined. This correction to the Z values is made on the next engine cycle. So when you watch the manifold convergence monitor for a single-cylinder configuration you will see that the masses past all four states become identical and equal to that through the valve. And if you run with a single cylinder on each manifold, you will see that these numbers become the same in both manifolds as the cylinder and manifold reach a stable periodic state.

TRICK FIVE

Now if you run a four-cylinder configuration you should see that the flow into the junction from the feeder is four times that into the computed runner, but it is not quite because the Z values assumed for the uncomputed runners are not quite right. Maybe you can guess what clever trick ESP uses to correct this problem. A mass balance on the junction is used to determine how much the Z values for the uncomputed runners need to be modified to bring the junction into proper mass balance, and this correction is applied on the next engine cycle. You can see this work when you watch the ESP manifold convergence monitor.

TRICK SIX

The time delays also need some help in forgetting their initial conditions so that they can become periodic as fast as possible. At the end of each cycle the time delays are reset to the average of the current value and the old average from previous cycles.

8.5

PROGRAM STATUS

The program was initially developed by W. C. Reynolds in the spring of 1987 without Section 8.4. At that time it was used by about seventy undergraduate students and six laboratory teaching assistants in the Mechanical Engineering Department at Stanford University. It was reviewed for a graduate class at Stanford in the fall of 1987, and refined slightly at that time. In 1997 changes were made in the burn stage of the program, making possible better agreement with actual engine data. The program was used again in the fall of 1997 in a Stanford undergraduate course, and in the spring of 1998 in Cornell and Ohio State courses. Between the spring of 1998 and the spring of 2000,[3] W. C. Reynolds developed and implemented ESP V2.0 for Windows and UNIX with the manifold model. This was first released for use in February 2000 with a command-line interface. A general purpose graphical user interface system, developed at Stanford by Stavros Kassinos and W. C. Reynolds, was being incorporated that spring. ESP is continually being improved, and suggestions are always welcome. See the README file on the ESP website for more details.

You are encouraged to use the model and report either problems or successes to the authors. See the README file for more information.

---

[3] The original ESP posted on the website in 1998 as this book first went to press was somehow damaged. This caused considerable anguish for users in the early spring of 2000, which was a time of intense work on the manifold version when support could no longer be provided for the old ESP. W. C. Reynolds apologizes to everyone for this problem, and hopes that those who were disappointed by bad experiences with the old version of ESP will try the new version to see what ESP can really do.

BIBLIOGRAPHY

[1] W. J. D. Annand and G. E. Roe. *Gas Flow in the Internal Combustion Engine.* Haessner Publisher Inc., Newfoundland, NJ, 1974.

[2] Anonymous. Jaguar Mark 2 Models, 2.4, 3.4 and 3.8 Litre, Service Manual. Technical report, Jaguar Cars (British Leyland UK Ltd.), Coventry CV5 9DR England, 1960. Publication Number E.121/7.

[3] Anonymous. Improving automobile fuel economy: new standards, new approaches. Technical report, Congress of the United States, Office of Technology Assessment, Washington, DC, 1991. For sale by the U.S. G.P.O., Supt. of Docs.

[4] Anonymous. Motor vehicle facts and figures. Technical report, Motor Vehicle Manufacturer's Association, 1992.

[5] Anonymous. Special feature: Active valve train system promises to eliminate camshafts. *Automotive Engineer*, pages 42–44, March 1993.

[6] Anonymous. *World Development Indicators.* World Bank, Washington, DC, 1997.

[7] C. Arcoumanis and J. H. Whitelaw. Fluid mechanics of internal combustion engines – a review. *Proceedings of the Institution of Mechanical Engineers*, 201(C1):57–74, 1987. See also ASME-FED-28.

[8] S. Arrhenius. Ueber die Reacktionsgeschwindigkeit bei der Inversion von Rohrzucher durch Saueren. *Zeitschrift fuer Physikalische Chemie*, 4:226–248, 1889.

[9] D. C. Haworth, S. H. El Tahry, and M. S. Huebler. A global approach to error estimation and physical diagnostics in multidimensional computational fluid dynamics. *International Journal for Numerical Methods in Fluids*, 17:75–97, 1993.

[10] E. M. Barber. Knock-limited performance of several automobile engines. *Transactions of the Society of Automotive Engineers*, 2(3):401–411, 433, July 1948.

[11] H. W. Barnes-Moss. A designer's viewpoint. In *Passenger Car engines, Conference Proceedings*, pages 133–147, London, 1975. Institution of Mechanical Engineers.

[12] G. K. Batchelor. *An introduction to fluid mechanics.* Cambridge University Press, London, 1967.

[13] Patrick Bedard. Cover story: Fast forward/Acura NSX-T, Dodge Viper gts, Porsche 911 turbo S. *Car and Driver*, pages 46–62, July 1997.

[14] L. Ben, J. Boree, and G. Charnay. Intake manifold pressure measurements in 2.2 L Renault J7T (4 cylinders; 2 valves/cylinder). Institut de Mécanique des Fluides de Toulouse; private communication, January 1999.

[15] M. Berckmueller, N. P. Tait, and D. A. Greenhalgh. The time-history of the mixture formation process in a lean burn stratified charge engine. Technical report, Society of Automotive Engineers, 1996. SAE 961929.

[16] Stuart Birch. Porsche 968. *Automotive Engineering*, 99(11):73–74, November 1991. Society of Automotive Engineers, Warrendale, PA.

[17] G. Borgeson. *The Classic Twin-Cam Engine*. Dalton Watson, Ltd., London, 1981.

[18] G. Borgeson. Twin cam – fine tuning the historical record. *Automobile Quarterly*, XXI(3):335, 1983.

[19] G. Borgeson. The 1914 grand prix Delage. *Automobile Quarterly*, XXIV(3): 306–318, 1986.

[20] G. Borgeson. The signature of the artist: debugging the desmodromic dichotomy and other historical fantasies. *Automobile Quarterly*, XXIV(3):319–323, 1986.

[21] J. D. Buckmaster and G. S. S. Ludford. *Theory of Laminar Flames*. Cambridge University Press, Cambridge, UK, 1982.

[22] C. E. Burke, L. H. Nagler, E. C. Campbell, L. C. Lundstrom, W. E. Zierer, H. L. Welch, T. D. Kosier, and W. A. McConnell. Where does all the power go? *SAE Transactions*, 65:713–737, 1957.

[23] W. H. Calvin. The great climate flip-flop. *The Atlantic Monthly*, 281(1):47–64, January 1998.

[24] Colin Campbell. *The Sports Car, Its Design and Performance*. Bentley Automotive Publishers, 1734 Massachusetts Avenue, Cambridge, MA 02138. (617) 547-4170. http://www.rb.com, 1959. 2nd Edition.

[25] W. K. Cheng, D. Hamrin, J. B. Heywood, S. Hochgreb, K. Min, and M. Morris. An overview of hydrocarbon emissions mechanisms in spark-ignition engines. Technical report, Society of Automotive Engineers, 1993. SAE 932708.

[26] J. M. Cherrie. Factors influencing valve temperatures in passenger car engines. Technical report, Society of Automotive Engineers, 1965. SAE No. 650484.

[27] R. Cogan. The fuel cell comes of age. *Motor Trend*, page 28, February 1998.

[28] Divers. *Guide to the Mercedes-Benz M-Class*. Road and Track, 1997. Special Series.

[29] F. V. Dolzhanskii, V. I. Klyatzkin, A. M. Obukhov, and M. A. Chusov. *Nonlinear Systems of Hydrodynamic Type*. Nauka, Moscow, 1974. In Russian.

[30] W. Demmelbauer-Ebner et al. Variable valve actuation systems for the optimization of engine torque. Technical report, Society of Automotive Engineers, 1991. SAE 910447.

[31] R. W. Fox and A. T. McDonald. *Introduction to Fluid Mechanics*. John Wiley & Sons, New York, etc., 1985.

[32] G. K. Fraidl, W. F. Piock, and M. Wirth. Gasoline direct injection: actual trends and future strategies for injection and combustion systems. Technical report, Society of Automotive Engineers, 1996. SAE 960456.

[33] S. Furuhama, M. Takiguchi, and K. Tomizawa. Effect of piston and piston ring designs on the piston friction forces in diesel engines. Technical report, Society of Automotive Engineers, 1981. SAE No. 810977; Trans. Vol. 90.

[34] N. E. Gallopoulos. Bridging the present to the future in personal transportation – the role of internal combustion engines. Reprint from: Fuel Systems and General Emissions (SP-910) GMR-7358, General Motors Research Laboratories, Warren, MI, February 1992. SAE 920721.

[35] Gamma Technologies, Inc., 601 Oakmont Lane, Suite 220, Westmont, IL 60559. *GT-Power*.

[36] Pat Ganahl. Roaring Subduer; the Greer-Black-Prudhomme Dragster. *The Rodder's Journal*, (Three):46–55, 1995. Spring.

[37] R. E. Gish, J. D. McCullough, J. B. Retzloff, and H. T. Mueller. Determination of true engine friction. *Transactions of the SAE*, 66:649–661, 1958.

[38] E. B. Gledzer, F. B. Dolzhanskii, and A. M. Obukhov. *Systems of hydrodynamic type and their applications*. Nauka, Moscow, 1981. In Russian.

[39] S. Goldstein, editor. *Modern Developments in Fluid Dynamics*. The Clarendon Press, Oxford, UK, 1952. First edition 1938; reprinted lithographically at the University Press, Oxford, 1943, 1950, 1952; in two volumes.

[40] D. L. Greene. Efficiency related changes in automobile and light truck markets 1978–1986. Technical report, Society of Automotive Engineers, 1986. SAE 861423.

[41] D. L. Greene and D. J. Santini. *Transportation and Global Climate Change*. American Council for an Energy Efficient Economy, Washington, D.C. and Berkeley, CA, 1993.

[42] The David Brown Group. *The Lagonda 2 1/2 Litre Instruction Book*. Lagonda Limited, Hanworth Park Works, Feltham, Middlesex, 1950.

[43] D. C. Haworth. Large eddy simulation of in-cylinder flows. In *Simulation of Engine Internal Flows; December 3–4, 1998*, Rueil-Malmaison, France, 1999. IFP. To be published.

[44] D. C. Haworth, S. H. El Tahry, and M. S. Huebler. Multidimensional port-and-cylinder flow calculations for two- and four-valve per cylinder engines: influence of intake configurations on flow structure. Technical report, Society of Automotive Engineers, 1990. SAE 900257.

[45] W. R. Hawthorne. Biographical memoir of H. R. Ricardo. In *Biographical Memoirs of the Royal Society*, volume 22, pages 359–380. Royal Society of London, London, UK, 1976.

[46] K. J. E. Hesselman. Internal combustion engine. United States Patent Office, Number 1,967,243, Oct. 29 1931.

[47] J. B. Heywood. *Internal Combustion Engine Fundamentals*. McGraw, 1988.

[48] M. Hochkoenig and M. Rauser. Cooling system layout for high performance cars. Technical report, Society of Automotive Engineers, 1992. SAE 920789.

[49] P. Holmes, J. L. Lumley, and G. Berkooz. *Turbulence, Coherent Structures, Dynamical Systems and Symmetry*. Cambridge University Press, Cambridge, UK, 1996.

[50] W.-H. Hucho and G. Sovran. Aerodynamics of road vehicles. *Annu. Rev. Fluid Mech.*, 25:485–537, 1993.

[51] J. C. R. Hunt. A review of the theory of rapidly distorted turbulent flow with its applications. *Fluid Dynamics Transactions*, 9:121–152, 1978.

[52] Gustav Ingmar Johnson. Studying valve dynamics with electronic computers. Technical report, Society of Automotive Engineers, 1963. SAE Paper number 596C; National Powerplant Meeting, Philadelphia, PA, Oct. 29–Nov. 2, 1962.

[53] G. B. Wood Jr., D. U. Hunter, E. S. Taylor, and C. F. Taylor. Air flow through intake valves. *SAE Trans.*, 50:212–220, 252, 1942.

[54] R. C. Juvinall and K. M. Marshek. *Fundamentals of Machine Component Design*. John Wiley & Sons, New York, etc., 1991.

[55] S. C. Kassinos and W. C. Reynolds. A structure-based model for the rapid-distortion of homogeneous turbulence. Technical report, Dept. of Mechanical Engineering, Stanford University, Stanford, CA., 1994. Ph.D. Thesis, Report TF-61s.

[56] I. W. Kay. Manifold fuel film effects in an IC engine. Technical report, Society of Automotive Engineers, 1978. SAE 780944.

[57] B. Khalighi, S. H. El Tahry, D. C. Haworth, and M. S. Huebler. Computation and measurement of flow and combustion in a four-valve engine with intake variations. Technical report, Society of Automotive Engineers, 1995. SAE 950287.

[58] T. Kim, S. Noh, C. Yu, and I. Kang. Optimization of swirl and tumble in KMC 2.4L lean burn engine. Technical report, Society of Automotive Engineers, 1994. SAE 940307.

[59] L. E. Kinsler, A. R. Frey, A. B. Coppens, and J. V. Sanders. *Fundamentals of Acoustics*. John Wiley and Son, New York, 1982.

[60] G. Konig and C. G. W. Sheppard. End gas autoignition and knock in a spark ignition engine. Technical report, Society of Automotive Engineers, 1990. SAE 902135.

[61] J. T. Kovach, E. A. Tsakiris, and L. T. Wong. Engine friction reduction for improved fuel economy. Technical report, Society of Automotive Engineers, 1982. SAE No. 820085.

[62] A. V. Kulkarni. New generation small block v8 engine. Technical report, Society of Automotive Engineers, 1992. SAE 920673.

[63] D. W. Lee. A study of air flow in an engine cylinder. Technical Report 653, NACA, Langley, VA, 1939.

[64] S. H. Lee, C. S. Jun, T. W. Roh, S. I. Kim, and H. S. Mok. The design and development of the new KIA T8D DOHC engine. Technical report, Society of Automotive Engineers, 1997. SAE 970917.

[65] J. C. Livengood and J. D. Stanitz. The effect of inlet valve design, size and lift on the air capacity and output of a four-stroke engine. Technical report, National Advisory Committee for Aeronautics, Washington, DC, 1943. TN 915; 32 pp incl. 18 figs.

[66] J. L. Lumley and H. A. Panofsky. *The Structure of Atmospheric Turbulence*. J. Wiley & Son., New York, 1964.

[67] F. Markus. Lean-burn engines. *Car and Driver*, pages 72–75, February 1992.

[68] F. Markus. A people's car: 25 horses, $6,000. *Car and Driver*, pages 133–138, December 1997.

[69] Ryoichi Matsumura, Kazuhiro Higashiyama, and Kazuo Kojima. The turbocharged 2.8L engine for the Datsun 280ZX. Technical report, Society of Automotive Engineers, 1982. SAE 820442.

[70] J. N. Mattavi and C. A. Amann, editors. *Combustion Modeling in Reciprocating Engines*. Plenum Press, New York, 1980.

[71] B. W. Millington and E. R. Hartles. Frictional losses in diesel engines. Technical report, Society of Automotive Engineers, 1968. SAE No. 680590.

[72] Y. Nakagawa, Y. Takagi, T. Itoh, and T. Iijima. Laser shadowgraphic analysis of knocking in S. I. engine. Technical report, Society of Automotive Engineers, 1984. SAE 845001.

[73] Y. Nakajima, T. Nagai, T. Iijima, and J. Yokayama. Analysis of combustion

patterns effective in improving anti-knock performance of a spark ignition engine. Technical report, Society of Automotive Engineers, 1984. JSAE Review, March.

[74] National Bureau of Standards, Washington, DC. *JANAF Thermochemical Tables*, 1971. Publication NSRDS-NBS37.

[75] Matt O'Keefe. Solar waxing. *Harvard Magazine*, 100(5):19–20, May–June 1998.

[76] Committee on Fuel Economy of Automobiles and Light Trucks. Automotive fuel economy – how far should we go? Technical report, Energy Engineering Board, Commission on Engineering and Technical Systems, National Research Council, 1992. Washington: National Academy Press.

[77] John Phillips. Oshkosh Phoenix. *Car and Driver*, pages 148–151, April 1997.

[78] S. G. Poulos and J. B. Heywood. The effect of chamber geometry on spark-ignition engine combustion. Technical report, The Society of Automotive Engineers, 1983. SAE 830334.

[79] D. L. Reuss. PIV velocity field measurements at TC in a motored 4-stroke 2-valve engine. GM Research & Development Center; private communication, January 1999.

[80] D. L Reuss, T.-W. Kuo, B. Khalighi, D. Haworth, and M. Rosalik. Particle image velocimetry measurements in a high-swirl engine used for evaluation of computational fluid dynamics calculations. Technical report, Society of Automotive Engineers, 1995. SAE 952381.

[81] W. C. Reynolds. Modeling of fluid motions in engines – an introductory overview. In J. N. Mattavi and C. A. Amann, editors, *Combustion Modeling in Reciprocating Engines*, pages 69–124. Plenum Press, 1980.

[82] Ricardo Software, Burr Ridge, IL. *WAVE Basic Manual: Documentation/ User's Manual*, October 1996. Version 3.4.

[83] J. R. Ristorcelli. Toward a turbulence constitutive relation for geophysical flows. In "New Directions in Geophysical Fluid Dynamics and Turbulence," *Theoret. Comput. Fluid Dynamics*, 9:207–221, 1997.

[84] J. R. Ristorcelli, J. L. Lumley, and R. Abid. A rapid-pressure covariance representation consistent with the Taylor-Proudman theorem materially-frame-indifferent in the 2D limit. *J. Fluid Mech.*, 292:111–152, 1995.

[85] R. C. Rosenberg. General friction considerations for engine design. Technical report, Society of Automotive Engineers, 1982. SAE 821576.

[86] A. M. Rothrock and R. C. Spencer. The influence of directed air flow on combustion in a spark-ignition engine. Technical Report 657, NACA, Langley, VA, 1939.

[87] S. Russ. A review of the effect of engine operating conditions on borderline knock. Technical report, Society of Automotive Engineers, 1996. SAE 960497.

[88] Ascher H. Shapiro. *The Dynamics and Thermodynamics of Compressible Fluid Flow*. The Ronald Press Company, New York, 1954.

[89] Paul Skilleter. 50 years of the XK engine: The pride of Lyons. *Thoroughbred and Classic Cars, The Pride and the Passion*, (296):66–70, May 1998. Peterborough, UK.

[90] U. Spicher and H. P. Kollmeier. Detection of flame propagation during knocking combustion by optical fiber diagnostics. Technical report, Society of Automotive Engineers, 1986. SAE 861532.

[91] U. Spicher and R. Krebs. Optical fiber technique as a tool to improve combustion efficiency. Technical report, Society of Automotive Engineers, 1990. SAE 902138.

[92] U. Spicher, H. Kroger, and J. Ganser. Detection of knocking combustion using simultaneously high-speed Schlieren cinematography and multi optical fiber technique. Technical report, Society of Automotive Engineers, 1991. SAE 912312.

[93] C. F. Taylor. Effect of size on the design and performance of internal combustion engines. *Transactions of the ASME*, July 1950.

[94] C. F. Taylor. *The Internal Combustion Engine in Theory and Practice*, Volume 1. M.I.T. Press, Cambridge, MA, 1966.

[95] C. F. Taylor. *The Internal Combustion Engine in Theory and Practice*, Volume 2. M.I.T. Press, Cambridge, MA, 1968.

[96] H. Tennekes and J. L. Lumley. *A First Course in Turbulence*. M.I.T. Press, Cambridge, MA, 1972.

[97] M. J. Tindal, T. J. Williams, and M. Aldoory. The effect of inlet port design on cylinder gas motion in direct injection diesel engines. In *Flows in Internal Combustion Engines*, pages 101–111. American Society of Mechanical Engineers, 1982.

[98] A. A. Townsend. *The Structure of Turbulent Shear Flow*. Cambridge University Press, Cambridge, UK, 1957.

[99] Eds U. Adler et al. *Automotive Handbook*. Robert Bosch GmbH, Stuttgart, 1993.

[100] T. Urushihara, T. Nakada, A. Kakuhou, and Y. Takagi. Effects of swirl/tumble motion on in-cylinder mixture formation in a lean-burn engine. Technical report, Society of Automotive Engineers, 1996. SAE 961994.

[101] V. A. Vladimirov and D. G. Vostretsov. Instability of steady flows with constant vorticity in vessels of elliptic cross-section. *Prikl. Matem. Mekhan.*, 50(J):369–377, 1986.

[102] M. L. Wald. In a step toward a better electric car, company uses fuel cell to get energy from gasoline. *New York Times*, 147:10(N), 14(L), Tuesday, October 21 1997. National Report Pages, column 1.

[103] Z. Warhaft. *An Introduction to Thermal-Fluid Engineering: The Engine and the Atmosphere*. Cambridge University Press, New York, 1997.

[104] Robert C. Weast, editor. *Handbook of Chemistry and Physics*. CRC Press, Cleveland, OH, 58th edition, 1977.

[105] N. D. Wilson, A. J. Watkins, and C. Dopson. Asymmetric valve strategies and their effect on combustion. Technical report, Society of Automotive Engineers, 1993. SAE 930821.

[106] L. Withrow and G. M. Rassweiler. Slow motion shows knocking and non-knocking explosions. *Transactions of the SAE*, 31:297–303, 312, 1936.

[107] F.-Q. Zhao, M.-C. Lai, and D. L. Harrington. A review of mixture preparation and combustion control strategies for spark-ignition direct-injection gasoline engines. Technical report, Society of Automotive Engineers, 1997. SAE 970627.

INDEX

Printed in the United States
By Bookmasters

Printed in the United States
By Bookmasters